Science/Technology/Society
as Reform in Science Education

SUNY Series in Science Education
Robert E. Yager, Editor

SCIENCE/TECHNOLOGY/SOCIETY
AS REFORM IN SCIENCE EDUCATION

Robert E. Yager, Editor

STATE UNIVERSITY OF NEW YORK PRESS

Published by
State University of New York Press, Albany

Printed in the United States of America

For information, address State University of New York
Press, State University Plaza, Albany, N.Y. 12246

Production by E. Moore
Marketing by Bernadette LaManna

Library of Congress Cataloging-in-Publication Data

Science/technology/society as reform in science education / edited by
Robert E. Yager.
p. cm. — (SUNY series in science education)
Includes bibliographical references and index.
ISBN 0-7914-2769-2 (hc acid-free). — ISBN 0-7914-2770-6 (pb acid
-free)
1. Science—Study and teaching—United States. 2. Technology-
-Study and teaching—United States. 3. Science and state—United
States. I. Yager, Robert Eugene, 1930- . II. Series.
Q183.3.A1S37 1996
507.1'073—dc20 95-10191
CIP

10 9 8 7 6 5 4 3 2 1

CONTENTS

PREFACE

Science/Technology/Society (STS) is widely recognized as a major reform effort as correctives are sought around the globe to attain a scientific literacy for all. Typical school science is ineffective in producing students who are knowledgeable of the basic laws and theories known to scientists as accepted views of the workings of nature. Nor are we successful in producing students who think that such views of the universe are important and/or relevant to their own lives.

Interestingly, technology (how the human-made world operates) is seen as more important today than science (how the natural world operates). And yet, it is rarely taught to all students across the elementary and middle school years. Each attempt at reform of science in the United States over the past 200 years has moved science instruction to the practical, that is, science that could affect the lives of graduates and tie to current technology. The only exception to this trend for reform was that characterizing the national efforts during the 1960s—following the Soviet launch of Sputnik in 1957. Although the artificial satellite was more of a technological achievement than a science one, attention and funding were directed toward reform that illustrated and emphasized basic science—that is, the science known and practiced by scientists. Scientists in their disciplines were instrumental in determining the central themes for physics, chemistry, biology, and earth science. The science disciplines were moved to the junior high school curriculum—a replication of the discipline bound programs of the high school. Process skills used by scientists became a central focus for science programs for the elementary school.

Reform was seen as a return to basic science, especially concepts and themes currently accepted by scientists. Reform also ushered in a renewed look at the process skills employed by scientists—an effort first popularized as a second dimension for the focus in school programs in the 1930s. Jerrold Zacharias, the architect of the Physical Science Curriculum Study (the one reform underway before 1957), identified the axiom that defined the science reforms of the 1960s and early 1970s. He said, "Science, when presented in a way known to scientists, will be inherently interesting and appropriate for all learners."

Disillusionment with these reforms following Sputnik was widespread by the mid-1970s. Many blamed the social problems, including the Vietnam conflict, on science and technology. Nearly all social institutions—including schools—were under attack.

Many efforts of the late 1970s and early 1980s resulted in a data base that was useful as new reforms were conceived. The economic woes in the U.S. hastened a fundamental change in United States policy that called for renewed efforts and funding for improving school science, technology, and mathematics. The appearance of technology was a major shift—something that is the connector for STS.

Many were still convinced that the efforts of the 1960s were correct and that we had merely abandoned them too soon. The controversy led to a focus on new research to find out more about the learning process. In 1983 funds were awarded to study physics and engineering majors at colleges and universities—with the simplistic notion that this information would help educators know how to deal with students in schools with less interest and expertise with science.

Surprisingly, however, the research found that these most interested and successful students had not learned. They could not use the information (concepts) they seemed to know in classes and on examinations. Nor could they use the process skills they "learned" and practiced in laboratories. It was soon apparent that the reform efforts for the 1980s would have to start in ways not generally tried or successful in the past.

This was the national situation as interest and trial with STS in science education was borne. The promise of STS was first seen in trials in other countries and circumstances. Such new foundations for reform provided the foundation for STS experiments across the United States. This volume is conceived as an effort to assemble research implications for the STS movement in the United States with some consideration to continuing initiatives the world over.

One of the major problems with any reform effort is visualizing all its features. STS efforts in the United States were so extensive by 1988 that LaMoine Motz, then president of the National Science Teachers Association (NSTA), appointed a Task Force to offer a definition of STS for NSTA—the world's largest professional society for science education. In 1991 the recommendations of this Task Force were unanimously adopted by NSTA. The essence of the definition is captured by the statement: STS is the teaching and *learning* of science-technology in the context of human experience. The report ended with the following statement: "The bottom line in STS is the involvement of learners in experiences and issues which are directly related to their lives.

STS develops students with skills which allow them to become active, responsible citizens by responding to issues which impact their lives. The experience of science education through STS strategies will create a scientifically literate citizenry for the twenty-first century" (NSTA, 1990–1991, p. 48).

In this volume the NSTA definition for STS is used. To be sure there are other definitions that are in use—definitions that focus on the curriculum. Too often they prescribe new lessons, new concepts, new activities that retain all the features of traditional textbooks and verification-type laboratories.

The focus of these twenty-nine chapters and five sections is upon STS offered as reform as a new century emerges. The efforts during the 1984–1994 decade have resulted in research that can be used to affect practices, while also assuring that STS deserves the designation as a reform effort. The first section is an elaboration of the meaning of STS and its ties to constructivism. The second section includes the chapters which summarize the research from classroom practice with STS. The third section deals with implications from the research which tell us what is needed if STS is to succeed. Section four relates efforts in the United States to those underway in the international arena. Section five provides information about lasting STS reform and indicates something of the future promise.

The work of all the authors indicated in the table of contents over a two-year period is acknowledged. All are convinced of the importance of sharing the research analyses and their implications as general reform of practice in schools and colleges remain illusive. Nonetheless, the new national standards in science (as developed by eighty-one professionals under the guidance of the National Research Council of the National Academy of Science) do exemplify STS as described in this monograph. Indeed, the standards may be possible because of the exciting experiences with STS and the research available attesting to its actual impact in stimulating real learning.

Reference

National Science Teachers Association (1990–1991). The NSTA position statement on science-technology-society (STS). In *NSTA handbook* (pp. 47–48). Washington, D.C.: author.

Part I

STS as a Reform Movement in Science Education

How has STS become a reform? What are its features? How does the definition of STS influence its practice and the emerging research results?

Since STS is primarily a focus on teaching and learning, the concern for the curriculum, materials for teachers to use, and similar artifacts for judging the nature of reform are more difficult to produce. Of course, there is a curriculum—one designed to help develop instructional goals. However, most reforms have failed because teaching has not changed.

Surely, the reforms of the 1960s were less successful because textbooks remained the determiners of courses (the textbooks were different when first produced but gradually were changed at the request of teachers to look like most textbooks before the interventions of the 1960s).

The focus on learning, that is, evidence for it and how teaching must change if we are to succeed with student learning, is central to STS instruction and the emerging research.

CHAPTER 1

HISTORY OF SCIENCE/TECHNOLOGY/SOCIETY AS REFORM IN THE UNITED STATES[1]

Robert E. Yager

SCIENCE IN THE SCHOOL CURRICULUM

Science has been an integral part of the school curriculum for our entire history in the United States. Basically, science has been a collection of courses in high school that reflect the major disciplines of science, that is, astronomy, botany, chemistry, geology (physical geography), physics, physiology, and zoology. Although science has enjoyed status as a core "subject" in the secondary school curriculum, along with language arts, mathematics, social studies, and foreign languages, it has never been considered as basic as language arts and mathematics, presumably because of the special skills characterizing these two curricular areas (quantification, measuring, reading, writing, and speaking). Nonetheless, science has been considered an important and at times a vital part of the kindergarten through twelfth-grade curriculum, especially during the past fifty years.

Unfortunately, high school science is invariably associated with preparation for college. And the courses prior to high school are thought to be preparatory for the next science course for the next academic year. Although there have been many reform efforts for school science over our 200+ year history, few have resulted in significant changes. Most courses have been organized around basic concepts—those identified as important in a state framework or those recognized as basic by various professional groups. The major determiner for science content has been standard science textbooks, where there has been found to be less than a 10 percent variation among those available for a given grade level (Harms and Yager, 1981).

THE STS MEGATREND

Science/Technology/Society has been called the current megatrend in science education (Roy, 1984). Others have described it as a paradigm shift for the field of science education (Hart and Robottom, 1990). In 1980 the National Science Teachers Association called STS the central goal for science education in its official Position Statement for the 1980s:

> The goal of science education during the 1980s is to develop scientifically literate individuals who understand how science, technology, and society influence one another and who are able to use their knowledge in their everyday decision-making. The scientifically literate person has a substantial knowledge base of facts, concepts, conceptual networks, and process skills which enable the individual to learn logically. This individual both appreciates the value of science and technology in society and understands their limitations. (NSTA, 1982, p. 1)

During the decade that followed, STS became the focus for two yearbooks for NSTA (Bybee, 1985; Bybee, Carlson, and McCormack, 1984) and one for the Association for the Education of Teachers of Science (James, 1986). STS sessions have become a program category for NSTA conventions. A new national organization has been formed—the National Association for Science, Technology, and Society (NASTS); it has a growing membership. There have been several major NSF grants awarded to foster STS approaches to school science and related curriculum fields. Two of the largest grants have been awarded to the Pennsylvania State University which boasts of establishing one of the first STS programs in a major U.S. university.

Rustum Roy of Penn State was the Principal Investigator of a major NSF grant in 1985, a project called Science through STS. The effort involved surveying all STS initiatives, kindergarten through college, throughout the United States and other nations. Materials were collected, a newsletter was initiated, and new instructional materials were developed. It was from these initiatives that NASTS was launched. A second grant established a network for promoting STS among science and social studies leaders in all fifty states; this network continues to provide a communication link among STS reformers.

Nearly every textbook publisher has embarked on actions to add STS materials in response to state mandates and local curriculum developments. Often industrial and private foundations have added support for specific STS projects. All indicators seem to suggest that STS indeed is a megatrend. How did it arise? How has it evolved? What is the rationale for the movement?

Origin

STS efforts were underway in several European countries before STS became a major focus in the United States. Two national programs have existed in the United Kingdom for several years; both are active and sponsored by the Association for Science Education in the United Kingdom. The first of these was Science in Society (Lewis, 1981) and the second is called Science in a Social Context (SISCON) (Solomon, 1983). Projekt Leerpakketontwikkeling Natuurkunde (PLON) is a well-established STS program in the Netherlands (Eijkelhof, Boeker, Raat, and Wijnbeek, 1981). *SciencePlus* is a curriculum development in Canada that enjoys widespread use in most provinces in the middle school years (ASCP, 1986, 1987, 1988).

STS as a term was coined by John Ziman in his book *Teaching and Learning About Science and Society* (1980). Ziman identified several courses and titles and special projects that had many common features. All were concerned with a view of science in a societal context—a kind of curriculum approach designed to make traditional concepts and processes found in typical science and social studies programs more appropriate and relevant to the lives of students.

STS in the United States

There have been many attempts in the United States to initiate STS programs in secondary schools. One such attempt centered at the University of Iowa in the Laboratory School in the early 1960s. Faculty from social studies and science conceived a course called Science and Culture, which met graduation requirements in science or in social studies. The course, in operation until the school closed in 1972, was funded by a grant from the Department of Education and was the subject of a PhD dissertation (Cossman, 1967) and some resulting publications (Yager and Casteel, 1966, 1968). The research indicated that students were able to attain and to retain many skills and competencies defined as science literacy. Such skills and competencies were not developed as a result of study in standard social studies or science courses.

Although the many efforts and their results were encouraging, STS did not get underway in the United States until 1981 with the report of Norris Harms's Project Synthesis study (1977). Harms included STS as one of five areas of concern as school science programs were studied in terms of how they met criteria for excellence established by expert task forces. Project Synthesis was organized around four goals clusters that served as one basis for a variety of analyses. These goal areas offered justifications for the inclusion of science in schools and for requiring it each year for ten to thirteen years. These four goal clusters were:

1. *Science for meeting personal needs*. Science education should prepare individuals to use science for improving their own lives and for coping with an increasingly technological world.

2. *Science for resolving current societal issues.* Science education should produce informed citizens prepared to deal responsibly with science-related societal issues.
3. *Science for assisting with career choices.* Science education should give all students an awareness of the nature and scope of a wide variety of science and technology-related careers open to students of varying aptitudes and interests.
4. *Science for preparing for further study.* Science education should allow students who are likely to pursue science academically as well as professionally to acquire the academic knowledge appropriate for their needs.

An analysis of the three National Science Foundation (NSF) status studies (Helgeson, Blosser, and Howe, 1977; Stake and Easley, 1978; Weiss, 1978) and the *Third Assessment of Science* by the National Assessment of Educational Progress (NAEP, 1978) were also basic parts of Harms's Project Synthesis. Several findings concerning the actual state of science teaching combined to encourage more attention to STS approaches. These included:

1. Ninety percent of all science teachers used textbooks for science instruction in excess of 90 percent of the time.
2. Textbooks were devoid of any considerations of the first three goal areas (material dealing with personal needs, societal issues, and/or career awareness).
3. Instruction focused on textbook readings, teacher lectures, question and answer techniques, and verification-type "laboratories."
4. Over 90 percent of the evaluation in science classes was based on the recall of information.
5. Teachers viewed themselves as the determiners of information to be covered and the evaluators for discovering the degree such information was acquired by each student.
6. The only goal area of concern to teachers and in evidence in schools was the fourth one, that is, preparing students for the further study of science.

Harms concluded his analysis of Project Synthesis report:

> . . . a new challenge for science education emerges. The question is this: "Can we shift our goals, programs, and practices from the current overwhelming emphasis on academic preparation for science careers for a few students to an emphasis on preparing all students to grapple successfully with science and technology in their own, everyday lives, as well as to participate knowledgeably in the important science-related decisions our country will have to make in the future?" (Harms and Yager, 1981, p. 119)

In one sense STS efforts are seen as responses to the first three goal clusters of Project Synthesis. STS means focusing on personal needs of students, that is, science concepts and process skills that are useful in the daily living of students. It focuses on societal issues, that is, issues and problems in homes, schools, and communities as well as the more global problems that should concern all humankind. STS also means focusing on the occupations and careers that are known today; it means using human resources in identifying and resolving local issues.

Evidence is mounting that concentrating on the first three goal clusters (STS foci) allows one to ignore goal area four. Students who are actively involved in studies that meet their personal needs, assist them to deal with current societal issues, and be aware of occupational-career possibilities, also find that science information is required—the same information that is widely accepted as needed preparation for further study in particular science disciplines. Students who experience their science in an STS format are well equipped to study and learn on their own, whether in college or in living outside of an educational institution.

For many, a focus on personal needs is an especially important concept for science in the elementary school. A focus on social issues and career awareness is often reserved for the middle and high school levels. However, when STS is viewed primarily as an approach to teaching and a meaningful view of science in people's lives, differences among the levels of teaching (i.e., kindergarten through college) become less significant than if STS is viewed primarily as a curriculum change.

STS is seen as a response to many of the perceived problems of traditional science teaching. The most critical problems with traditional science teaching are:

1. Students generally cannot use the science (either concepts or processes) that they learn. The number of misconceptions that typical high school students have is large. Misconceptions that the most successful students have are shocking. For example, recent reports indicate that 80 percent of university physics majors have misconceptions about nature even though they recite correct factual information and can perform exercises in the laboratory that contradict their own views of the world (Champagne and Klopfer, 1984). As many as 90 percent of engineering majors cannot relate their preparation to the real-world (Mestre and Lochhead, 1990).
2. Well over 90 percent of all high school graduates do not attain scientific literacy—even though they pass courses and generally perform well (Miller, 1989; Miller, Suchner, and Voelker, 1980). Science instruction does not seem to produce persons with traits of scientific literacy that are deemed important—perhaps the fundamental goal of instruction (see quote from 1982 NSTA Position Statement in the opening paragraph).

3. Interest in science and initiation of further study of science declines across the K–12 years. In fact, positive attitudes about science, science classes, science teachers, and the usefulness of science to living decline the more science is studied in school (ETS, 1988; Hueftle, Rakow, and Welch, 1983; NAEP, 1978; Yager and Penick, 1986).
4. Creativity is central to basic science, including the questions asked of nature, the explanations offered, and the tests devised to determine the validity of such explanations. And yet the study of typical science results in a diminution of the creativity skills originally possessed. Typical science instruction causes students to be less curious, less prone to offer explanations, less able to devise tests, less able to predict causes and consequences of certain actions (ETS, 1988; Hueftle, Rakow, and Welch, 1983; NAEP, 1978; Yager and Penick, 1986).
5. There is no evidence that traditional science teaching results in persons who possess the traits which characterize a scientifically literate person. NSTA adopted a listing of the characteristics of a person who is scientifically literate. Such a person:
   a. uses concepts of science and of technology and ethical values in solving everyday problems and making responsible everyday decisions in everyday life, including work and leisure;
   b. engages in responsible personal and civic actions after weighing the possible consequences of alternative options;
   c. defends decisions and actions using rational arguments based on evidence;
   d. engages in science and technology for the excitement and the explanations they provide;
   e. displays curiosity about and appreciation of the natural and human-made world;
   f. applies skepticism, careful methods, logical reasoning, and creativity in investigating the observable universe;
   g. values scientific research and technological problem solving;
   h. locates, collects, analyzes, and evaluates sources of scientific and technological information and uses these sources in solving problems, making decisions, and taking actions;
   i. distinguishes between scientific-technological evidence and personal opinion and between reliable and unreliable information;
   j. remains open to new evidence and the tentativeness of scientific-technological knowledge;
   k. recognizes that science and technology are human endeavors;
   l. weighs the benefits and burdens of scientific and technological development;
   m. recognizes the strengths and limitations of science and technology for advancing human welfare;

n. analyzes interactions among science, technology, and society;
o. connects science and technology to other human endeavors, for example, history, mathematics, the arts, and the humanities;
p. considers the political, economic, moral, and ethical aspects of science and technology as they related to personal and global issues; and
q. offers explanations of natural phenomena that may be tested for their validity (NSTA, 1990).

STS means viewing science in a way quite different from the post-Sputnik period where the emphasis was on the identification of the central concepts, the unifying themes, and/or the major theories that characterize the various science disciplines if not science itself. The prevailing view is that science could be made meaningful, exciting, and appropriate for all if it were presented in a way known to scientists. Science educators were anxious to see, learn, and transmit this view of science to students. There was no chance for student ownership, student questions, or student views of the world in which they lived. Rather, the attempt was to get students into the world seen, known, and experienced by scientists; that was identified as the major task of the science teacher.

During the 1960s every effort was made to distinguish between science and technology. Basic science was a focus and technology was stricken from courses labeled "science"! STS means using technology as a connector between science and society. The applications of science are seen as closer to the lives of students, including advances and issues concerning food, clothing, shelter, transportation, communication, and careers.

Certainly STS is viewing school science in broader terms than merely the science concepts accepted by practicing scientists and the process skills they use to discover new concepts and/or to test old ones. STS assumes that equating science only to specific concepts and processes and then assessing the degree each has been acquired is not an adequate indicator of real learning. Such practices provide no information concerning how the concepts and processes can be used in the lives of students and for future problem resolution.

STS as a Means for Meeting Educational Goals

If STS is proclaimed a megatrend in science education, it must focus on educational goals and unifying themes that tie most disciplines together to meet common goals. The strength of STS is the use of personal, societal, and career imperatives as organizers for curriculum. Such organizers bring relevance to study and build on past and continuing experiences of students. STS, when considered broadly, is free of specific topics, its own concepts, special processes, and unique teaching strategies. In final analysis STS is focusing on real issues of today with the belief that working on them will require the concepts and processes so many consider basic. In traditional schools and curriculum outlines, the concepts and processes of a given discipline are central. Time and effort are expended to figure out better ways to *present* this information and these skills to students. STS means starting with a situation—a question, problem, or issue—where a creative teacher can help students see the power and utility of basic concepts and processes. STS means starting with students and their questions, using all resources available to work for problem resolution, and, whenever possible, advancing to the stage of taking actual actions individually and in groups to resolve actual issues. STS makes science instruction current and a part of the real world. STS provides a context for learning basic concepts and process skills.

STS means dealing with students in their own environments and with their own frames of reference. It means moving into the world of applications, the world of technology, the world where the student makes his or her own connections to living and to the traditional disciplines.

Dealing with the real-world and problems in it tends to improve student attitudes and to use and sharpen creativity skills. These are called the enabling domains. They provide access to the concepts and processes as seen, advanced, and practiced by the professionals in a given discipline. When one starts with these concepts and processes (as in the case in traditional discipline-bound programs), most students are lost before they can apply anything to their own lives. Attitude worsens and creativity skills decline as one considers concepts and processes for their own merit and centrality. Those who maintain that scientific literacy is a nongoal usually assume that such literacy is dependent on the mastery of such standard concepts and processes. They insist that it is impossible to make all students knowledgeable of all basic/central concepts and processes that characterize a discipline. And this is so—if one accepts a definition of science/technological literacy and focuses only on a recitation of "basic" concepts and process skills.

Concept mastery is a goal. But for mastery to exemplify real learning, information and process skills must be demonstrated as useful. Such a situation seldom occurs as a result of typical instruction. STS means that concepts and processes are useful because they are encountered when the student needs them to deal with problems he or she identifies. This occurs because of high motivation and interest and because the student has formulated questions, has offered explanations, and is interested in the validity of these explanations. This is science and these are basic ingredients of creativity.

REQUIREMENTS FOR IMPLEMENTING STS

STS teaching requires new models for pre- and inservice teacher education. One of the greatest problems associated with shifts to STS teaching is the failure of most teachers, even those newly certified, to have ever experienced science study and learning themselves as STS, that is, learning in the context of human experience. The current focus upon the Constructivist Learning Model (Yager, 1991; Yeany, 1990) indicates the importance of learning (including learning to teach differently) by direct personal experience.

A rationale/framework for STS can be discerned from a set of contrasts dealing with concepts, processes, attitudes, creativity skills, and applications. Tables 1.1–1.5 provide lists of these contrasts.

TABLE 1.1 Contrasts of Student Mastery of Concepts Emerging from Traditional and STS Classes

Traditional	*STS*
1. Concepts are really bits of information mastered for a teacher test	1. Students see concepts as personally useful
2. Concepts are seen as outcome themselves	2. Concepts are seen as a needed commodity for dealing with the problems
3. "Learning" is principally for testing	3. Learning occurs because of activity; it is an important happening but not a focus in and of itself
4. Retention is very short lived	4. Students who learn by experience retain it and can often relate it to new situations.

TABLE 1.2 Contrasts of Student Process Skills Emerging from Traditional and STS Classes

Traditional	*STS*
1. Students see science processes as skills scientists possess	1. Students see science processes as skills they can use
2. Students see processes as something to practice as a course requirement	2. Students see processes as skills they need to refine and develop more fully for themselves
3. Teacher concerns for process are not understood by students, especially since they rarely affect course grades	3. Students readily see the relationship of science processes to their own actions
4. Students see science processes as abstract, glorified, unattainable skills unrelated to their lives	4. Students see processes as vital parts of what they do in science classes

TABLE 1.3 Contrasts of Student Attitudes Emerging from Traditional and STS Classes

Traditional	*STS*
1. Student interest declines at a particular grade level and across grade levels	1. Student interest increases in specific courses and from grade to grade
2. Science seems to decrease curiosity about the natural world	2. Students become more curious about the natural world
3. Students see teacher as a purveyor of information	3. Students see teacher as a facilitator/guide
4. Students see science as information to learn	4. Students see science as a way of dealing with problems

TABLE 1.4 Contrasts of Student Creativity Skills Emerging from Traditional and STS Classes

Traditional	*STS*
1. Students decline in their ability to question; the questions they do raise are often ignored because they do not fit into the course outline	1. Students ask more questions; such questions are used to plan activities and use materials
2. Students rarely ask unique questions	2. Students frequently ask unique questions that excite their own interests, that of other students, and that of the teacher
3. Students are ineffective in identifying possible causes and possible effects of specific situations	3. Students have skills needed to suggest possible causes and effects of certain observations and actions
4. Students have few original ideas	4. Students seem to effervesce with ideas

TABLE 1.5 Contrasts of Application of Science Concepts Emerging from Traditional and STS Classes

Traditional	*STS*
1. Students see no value and/or use of their science study to their living	1. Students can relate their science study to their daily living
2. Students see no value in their science study for resolving current societal problems	2. Students become involved in resolving social issues; they see the relativity of science study to fulfilling citizenship responsibilities
3. Students can recite information/concepts studied	3. Students seek out information to use in dealing with questions
4. Students cannot relate the science they study to any current technology	4. Students are engrossed in current technological developments and use them to see the importance and relevance of science concepts

STS as a movement is less than ten years old in the United States. In that short time it has grown from a seemingly new idea to a major effort in every state. There remain conflicts as to what it is and what it is not. Many cannot deal with a movement like STS, which is not curriculum based. Instead of a curriculum it is a context for a curriculum. Many want to reserve judgment on STS until they see a curriculum and some goals and assessment instruments focused on basic concepts. Others are moving from STS to integrated science themes thereby retaining a more common concept of science courses and topics in them. Many in the STS movement are resisting the temptations of preparing a curriculum outline, of adding STS strands to existing courses and textbooks, of identifying new lists of concepts and processes, or preparing new examinations to assess the degree of recall of the new concepts and process skills. They even resist the temptation to move to identifying the effort as one of integrating science concepts from a variety of disciplines. To provide this framework can mean the end of Roy's Megatrend and/or Hart and Robottom's suggestion that STS represents a Paradigm Shift.

NOTE

1. A version of this article appeared in *School Science and Mathematics*, 1993, Volume 93, Issue 3, pp. 145-151.

REFERENCES

Atlantic Science Curriculum Project. (1986). *SciencePlus 1*. Toronto: Harcourt Brace Jovanovich, Canada.

Atlantic Science Curriculum Project. (1987). *SciencePlus 2*. Toronto: Harcourt Brace Jovanovich, Canada.

Atlantic Science Curriculum Project. (1988). *SciencePlus 3*. Toronto: Harcourt Brace Jovanovich, Canada.

Bybee, R. W. (ed.). (1985). *NSTA yearbook: Science/technology/society*. Washington, D.C.: National Science Teachers Association.

Bybee, R. W., Carlson, J., and McCormack, A. J. (eds.). (1984). *NSTA yearbook: Redesigning science and technology education*. Washington, D.C.: National Science Teachers Association.

Champagne, A. B., and Klopfer, L. E. (1984). Research in science education: The cognitive psychology perspective. In D. Holdzkom and P. B. Lutz, (eds.), *Research within reach: Science education* (pp. 171–189). Charleston, W.V.: Research and Development Interpretation Service, Appalachia Educational Laboratory.

Cossman, G. W. (1967). *The effects of a course in science and culture designed for secondary school students*. Unpublished doctoral dissertation, University of Iowa, Iowa City, Iowa.

Educational Testing Service. (1988). *The science report card: Trends and achievement based on the 1986 national assessment* (Rep. No. 17-S-01). Princeton: National Assessment of Educational Progress.

Eijkelhof, H. M. C., Boeker, E., Raat, J. H., and Wijnbeek, N. J. (1981). *Physics in society*. Amsterdam: VU-Bookshop.

Harms, N. C. (1977). Project Synthesis: An interpretative consolidation of research identifying needs in natural science education. (A proposal prepared for the National Science Foundation.) Boulder: University of Colorado.

Harms, N. C., and Yager, R. E. (eds.). (1981). *What research says to the science teacher, Vol. 3*. Washington, D.C.: National Science Teachers Association.

Hart, E. P., and Robottom, I. M. (1990). The science-technology-society movement in science education: A critique of the reform process. *Journal of Research in Science Teaching*, *27*(6), 575–588.

Helgeson, S. L., Blosser, P. E., and Howe, R. W. (1977). *The status of pre-college science, mathematics, and social science education: 1955–75*. Columbus, Ohio: Center for Science and Mathematics Education, The Ohio State University.

Hueftle, S. J., Rakow, S. J., and Welch, W. W. (1983). *Images of science: A summary of results from the 1981–82 national assessment in science*. Minneapolis: Minnesota Research and Evaluation Center, The University of Minnesota.

James, R. K. (ed.). (1986). *1985 AETS yearbook—Science, technology and society: Resources for science educators*. Columbus, Ohio: SMEAC Information Reference Center and Association for the Education of Teachers in Science.

Lewis, J. (1981). *Science and society*. London, England: Heinemann Educational Books and Association for Science Education.

Mestre, J. P., and Lochhead, J. (1990). *Academic preparation in science: Teaching for transition from high school to college*. New York, N.Y.: College Entrance Examination Board.

Miller, J. (1989, April). *Scientific literacy*. Paper presented at the meeting of the American Association for the Advancement of Science, San Francisco, California.

Miller, J. D., Suchner, R. W., and Voelker, A. M. (1980). *Citizenship in an age of science: Changing attitudes among young adults*. New York: Permagon Press.

National Assessment of Educational Progress. (1978). *The third assessment of science, 1976–77*. Denver: Author.

National Science Teachers Association. (1982). *Science-technology-society: Science education for the 1980s*. Position Paper. Washington, D.C.: Author.

National Science Teachers Association. (1990). Science/technology/society: A new effort for providing appropriate science for all (The NSTA position statement). *Bulletin of Science, Technology and Society, 10*(5&6), 249–250.

Roy, R. (1984). *S-S/T/S project: Teaching science via science, technology, and society material in the pre-college years.* University Park, Pa.: The Pennsylvania State University.

Solomon, J. (1983). *Science in a social context (SisCon).* United Kingdom: Basil Blackwell and the Association for Science Education.

Stake, R. E., and Easley, J. (1978). *Case studies in science education,* Vols. I and II (Stock No. 038-000-00364). Washington, D.C.: U. S. Government Printing Office, Center for Instructional Research and Curriculum Evaluation, University of Illinois at Urbana-Champaign.

Weiss, I. R. (1978). *Report of the 1977 national survey of science, mathematics, and social studies education: Center for educational research and evaluation.* Washington, D.C.: U.S. Government Printing Office.

Yager, R. E. (1991). The constructivist learning model: Towards real reform in science education. *The Science Teacher, 58*(6), 52–57.

Yager, R. E., and Casteel, J. D. (1966). Science as mind-affected and mind-affecting inquiry. *Journal of Research in Science Teaching, 4,* 127–136.

Yager, R. E., and Casteel, J. D. (1968). The University of Iowa science and culture project. *School Science and Mathematics, 67*(5), 412–416.

Yager, R. E., and Penick, J. E. (1986). Perceptions of four age groups toward science classes, teachers, and the value of science. *Science Education, 70*(4), 355–363.

Yeany, R. (1990, April). *Response to von Glasersfeld.* Paper presented at the general session of the meeting of the National Association for Research in Science Teaching, Atlanta.

Ziman, J. (1980). *Teaching and learning about science and society.* Cambridge: Cambridge University Press.

CHAPTER 2

MEANING OF STS FOR SCIENCE TEACHERS

Robert E. Yager

THE MEANING OF SCIENCE

Science to most people means studying about: (1) living forms, stars and planets, chemicals, the earth, mechanics, energy, and (2) other more specific topics found in textbooks about science and/or the major disciplines of science, namely, biology, chemistry, physics, and earth science. Such topics of science are fascinating to some persons—often over half of the students in the elementary school—but the number decreases to only a fourth of the students in high schools (NAEP, 1978; Weiss, 1987). Most adults look back on their experiences with school science in negative ways—at least in terms of what most science teachers advance as their objectives and intentions for their courses and their instruction (Yager, 1985). Most people identify science as the material studied in science classes; that is, the science that they report to be not too useful, nor particularly valuable, nor meaningful to their day-to-day pursuits. And yet these same adults feel that such experiences with school science are valuable for their own children to have. Over 85 percent of the adults today support rigorous experience with school science for their children—even when they cannot identify the value of their own experiences with it when they were students.

This situation is an intriguing one. Many science educators are convinced that a major problem is the lack of any utilitarian (useful) definition of science; most people never think beyond their own experience with studying science in school. And yet, what is typically studied represents only one dimension of science—perhaps even an unimportant dimension. Most persons—when pressed to offer a short, simple, concise definition of science—focus on the informational dimension. For example, Campbell (1957) views science to be the generalizations concerning the natural world about which experts agree. Feynman defines science to be basic understandings of the universe and its contents (Feynman, 1985). Such definitions suggest a certain structure, a certain information base, and a certain degree of comprehension for students to master. Such definitions limit learning by concentrating on information that is presumed to be prerequisite to real understanding. Unfortunately, such unidimensional definitions allow teachers, curriculum developers, and the general public to view science as only the mastery of certain concepts, those identified by contemporary scientists.

George Gaylord Simpson, biologist/paleontologist, has advanced a definition of science that is simple, easily associated with living, and yet more complete and comprehensive in describing what science is in practice (Pittendrigh and Tiffany, 1957). It is one that most scientists accept and one that is particularly attractive to science educators. Simpson's definition has three parts, suggesting three necessary ingredients for any human enterprise to be called "science." Simpson's definition is: Science is an exploration of the material universe in order to seek orderly explanations (generalizable knowledge) of the objects and events encountered: *but these explanations must be testable* (Simpson, 1963; Brandwein, 1983). Simpson's definition restricts the domain of science to the material universe. It identifies the major action to be one of explaining (understanding); it is a personal act, a creative act, an act requiring skills that presumably can be improved and sharpened. But such explanations must be testable if they are to be called science. Simpson recognizes that there are many acceptable explanations for phenomena, ideas, and human perception; however, if a test of such explanations cannot be devised, one cannot call the activity science. There are fine explanations of objects and events that are appropriate for fine arts, humanities, religion, and philosophy. Simpson's view of science enables us to see more dimensions of science, to see more of the total human enterprise, to consider more basic ingredients than normally found in the study of science in schools and colleges. It provides a richer context to consider science and its place among the various facets of the total curriculum for all learners.

THE MEANING OF SCIENCE TEACHING: GOALS AND PROBLEMS

Many problems exist in schools with respect to meeting instructional goals. Many of these problems could be reduced if educators would use Simpson's definition of science; more of the essential ingredients of science would be practiced and incorporated into the fabric of the curriculum and the teaching strategies employed. At present most teachers argue that they are aiding students in "understanding" science by reviewing the information found in standard textbooks—textbooks that seem to focus on the structure of the various disciplines and the generalizations currently accepted by professionals. Often these generalizations are abstract and important and/or attractive only to practicing scientists. Infrequently is there any attempt to consider generalities that are useful and/or important in the lives of people.

Generally the real problems in school science can be summarized with the following statements:

1. The science textbook is the source of nearly all information successful students need to learn; it is the source of teacher questions, questions for quizzes and examinations, and ideas for activities.
2. Most information in science courses is justified as necessary before one enrolls in the next course; rarely, however, is this the actual case. Most teachers present, consider, and review everything that students in a given course at a given grade level will need to perform in an exemplary manner.
3. The impact of science information on the lives of students and any applications of the concepts for students and/or society in general are omitted. It is assumed that impact and application will occur "naturally" or that other teachers in other curriculum areas will tend to this need.
4. Teachers view themselves as sources of information that is important for students to learn. They rarely admit to not knowing; they restrict student interest and attention to a rigid course outline.
5. Evaluation is based on vocabulary acquisition and recall of information from textbooks and/or teacher lectures.
6. Science is restricted to what occurs in a science classroom. There is rarely any emphasis on extensions of activities beyond the classroom or the school; it is rare to depend on resources of any kind beyond the textbook, the teacher, and the science room.

The Meaning of STS

Central to the STS approach is the focus on individual learners. Certainly everyone is a part of a society; the society (social structure) for younger students is more closely tied to the home, community, school, and classroom than is the society for older students, who can conceptualize a global society. The STS approach is one that necessitates problem identification by individual students and individual classes. Such problem identification includes—by its very nature—a multidisciplinary view. There are few problems that are related only to science—certainly not to one science discipline. For most people, societal problems by their very nature exist in a total context.

Science is habitually taught as if all students could and should become practicing scientists. Teachers often give special attention and praise to outstanding students in the traditional school setting. Teachers frequently brag about the most gifted students, taking special note of their success in college and their choice of professional science, medicine, or engineering careers. And yet, the number from a given high school graduating class who achieve such success is less than 1 percent of the total. Leyden has observed that the typical high school graduating class will contain more criminals than scientists (Leyden, 1984).

In addition to beginning with society—as defined by learners at given grade levels—the STS approach includes technology. Many science teachers find this to be a problem since most of their formal preparation was devoid of any work in technology (applied fields such as medicine, allied health, homemaking, industrial arts, engineering, agriculture, forestry). Science teachers are prepared in the pure sciences, taking college courses that are structured exactly like high school courses. If one were to examine the tables of contents of beginning texts in biology, chemistry, physics, or earth science, it would be difficult to unearth variations between high school and college tomes.

Most science teachers find it difficult to start with society and move to technology. After all, they are among the few who have succeeded with school science and who have continued with a bachelor's degree in science. They are the ones who find it most difficult to view science in any way other than how they have experienced it. This usually does not include any exploration for the sheer enjoyment of it. It does not usually allow students to formulate any explanations of their own; it certainly does not include any experience with devising and carrying out tests on the explanations advanced. The typical high school and college experience would not include any of the ingredients of basic science as envisioned and defined by Simpson.

Focusing first on questions and/or issues is appropriate for most people. Such a focus is basic to reports in newspapers, magazines, and television. Issues comprise the major treatment for popular science publications such as *Discover*, *Omni*, and *WorldWatch*. Most people are fascinated with the unknown, genuine problems, and complex issues. Few recognize such actions as decision-making, probing, debate, and conflict as having any relationship to science. And yet, in excess of 90 percent of all societal issues today are grounded in science/technology and require such actions.

Using societal issues as organizers for school science provides several advantages over the organization most recognize and the one that most teachers prefer. First of all, the use of issues provides a ready vehicle for utilizing the more complete definition of science as advanced by Simpson. The issue, the question, the uncertainty become the point of departure and the reason for exploring. Issues by their very definition are motivating, thought provoking, calls-to-action. There is a need for information and a reason for offering explanations. There is opportunity for much human interaction, including discussions of the relative validity of various explanations, the search for information-knowledge, and the formulation of tests for various explanations. Instead of students being told they need to have information (as in the typical classroom), students are the ones who recognize the need for information; they seek it out, apply it, and use it. Suddenly the problems most science teachers encounter in terms of motivation are solved.

The use of issues as organizers for school science also resolves the major problems of science education as elaborated previously. Suddenly the textbook is not "the course"; it is no longer necessary to preach about the value of science understanding. *Use* of science information is suddenly the beginning point rather than a difficult "add on"; the teacher's role is that of facilitator and helper, instead of dispenser of information; student success can be observed in performance terms, including application and synthesis, as opposed to recall. Science is seen as something that occurs everywhere—not just in books and science classes.

The STS approach provides other advantages—perhaps even more important than providing a better experience with real (and more complete) science and resolving the basic problems observed in science education circles. Such an approach may provide a major vehicle for tying the whole school program together. Using actual community, regional, and global problems provides obvious ties to all aspects of the school program. It can provide a means for the school to become a microcosm of society as a whole—allowing students to practice directly the skills needed for living in the adult world. Instead of assuming that all parts of the school program will be meaningful and that the parts over time will be useful and appropriate, student needs and experiences demonstrate the value and power of science explanation skills.

Using current issues as organizers for instruction is not foreign to other curriculum areas. Nonetheless, it is not a common pattern. Social studies teachers and their traditional courses of study are too often tied to discipline areas such as world history, American history, American government, sociology, economics, geography, and psychology. Any ties to world problems and any ties to science (as commonly approached in schools) are rarely observed. Many social studies teachers fear dealing with current issues—often exactly because of the ties to science and technology. Traditionally, social studies teachers feel threatened by these fields, inadequate to deal with such issues because of inadequate educational backgrounds.

Language arts teachers experience some of the same problems. They can deal with the mechanics of reading, writing, and speaking; they can teach journalism, drama, speech, debate, and literature. It is easy to focus on the concrete and the mechanics. All too often, however, that which is concrete is not relevant. All too often the mechanics seem to have no value in the real world. Mathematics is often a problem because it is taught as specialized skills for which applications in daily life are not apparent. An STS approach can focus on use of mathematics skills in a real-world context.

The Value of a Relevant Curriculum

Dewey (1938) often talked and wrote about the value of a relevant curriculum—the necessity for real learning to come from actual involvement. The arbitrary division of the curriculum and the school day into subjects/courses may be detrimental to real learning. We may be missing every opportunity to involve students directly in real learning, and missing the chance of offering a program that is preparatory for students who will live and operate in a future not totally unlike the present. Instead of isolating students in places called schools and forcing them to learn information from books and teachers—interactions that often seem unrelated to the world today—perhaps a focus on the world today, that is, current issues and problems, will provide the missing ingredient that is necessary for real learning to occur.

Cognitive psychologists (Champagne and Klopfer, 1984) are united in their observations that most high school graduates (college students and adults as well) have many naive theories (misconceptions) about the real world. They seem to retain these views even when they complete and succeed in advanced courses dealing with scientific concepts and theories. Apparently, the science learned in school is not internalized, that is, not really learned. The best students perform well on tests and advance to the next level; however, they internalize and believe interpretations of the real world based on real experiences they have had. There is often a discrepancy between school science and real-world experience. The real-world experience wins out every time over the science of textbooks, classrooms/laboratories, and school. There is every reason to believe that similar discrepancies exist between school experiences and personal experiences for students with regard to all other facets of typical school programs.

There is little worry about a world without issues. Some fear that more issues—and more complicated ones—arise than can be dealt with. There are problems of energy depletion, population explosion, food supplies, toxic wastes, nuclear proliferation, nutrition, disease, warfare, communication, transportation, agricultural production, synthetics, computerization, and space exploration. It is difficult to conceive of any newsworthy current event that is not a science-technology-related issue. Appropriate issues will provide needs for mathematics and a point of departure for social studies. As for language arts, why should not current problems and issues be used to practice good communication, and to develop better writing, reading, and speaking skills? It would seem that an STS focus in a school would have advantages for all students—if they view school as preparation for living. Further, most curriculum areas could meet their objectives more easily since students would be motivated to learn because they would see a need. Teachers would not have to insist that their students "need to know" and yet find it difficult to justify the need.

Some schools, such as those in Chariton, Iowa, involve students and teachers in selecting an annual STS issue. For two years space technology has been chosen. Every department gets involved with the issue. Many aspects of space exploration are rooted in basic physical science. Industrial technology is directly involved with communication systems, principles of flight, and design and construction. There are economic, psychological, government, and sociological aspects—that is, the social science focus. The need for communication, that is, letter writing, public forums, debate, and journalism is great. The art department is intimately involved; there are also direct ties to foreign language. Mathematics is needed in a variety of ways—always because of the need—not because there is another important skill to teach.

Many STS projects start with a common—and at times a comical situation. Some have included clogged toilets, a dripping faucet, a disaster caused by severe weather, an accident, or a news event arising from the Persian Gulf war. It is often amazing to see the connections students can make to basic science, and to the basic content of a variety of school disciplines, when they are encouraged to do so. When content is introduced by the teacher, textbook, or course of study, it is rare for the students (or teacher) to see or to identify any connectors to daily lives.

The Meaning of STS for Science Teachers

In every instance effective STS programs can be observed to affect teachers and learners more than did science alone. The focus on community and world problems becomes something with which all can identify. The community becomes the laboratory—not just a place called a science laboratory. The amount of communication among the faculty as a whole increases by 50 percent. Administrators—central office as well as building—become pivotally involved. In some instances, administrators are directly involved with establishing the new programs. In all instances community leaders become directly involved in the school program. New alliances are created; new partnerships are formed; in several schools community members form new organizations to support school efforts. Teacher inservice becomes much more important; teachers are quick to realize that they must continue to grow: the best teachers are involved learners. In all cases teachers become much more involved with each other, with students, with school officials, with community leaders, and with persons across the nation and world.

While real societal issues may provide a significant organizing force for the school program, expertise with developing needed communication skills is a vital link in developing an exemplary school program. Skills with interviewing, writing, speaking, reading, and debating are needed if teachers and students are to become involved in resolving school, community, state, national, and international problems. To be sure, most science teachers have too few skills in dealing with societal issues (traditionally viewed as the realm of the social studies), in dealing with applications of science (traditionally handled in applied areas such as agriculture, industrial arts, and homemaking), and in dealing with communication (the primary responsibility of educators in language arts). A focus on local issues can help schoolwork to be seen as more relevant, as teachers and students work together on one or more problems. Such cooperative efforts cannot help but to serve as a better model for future citizens who must work together to resolve problems and to advance our common culture.

STS has been introduced as an organizer for curricula in some states. Statewide inservice has provided a focus in such states as Arizona, New York, Iowa, New Jersey, and Wisconsin. Most science and social studies textbooks are introducing STS themes. However, the richness is not found in such new topics in textbooks, or in new units prepared by teachers. The richness of STS comes from students, and the central role they play in planning and carrying out their problem solving activities. STS often results in restructuring the total school program.

REFERENCES

Brandwein, P. F. (1983). *Notes toward a renewal in the teaching of science*. Chicago, Ill.: Coronado Publishers.

Campbell, N. R. (1957). *Foundations of science*. New York, N.Y.: Dover Publications, pp. 215–228.

Champagne, A. B., and Klopfer, L. E. (1984). Research in science education: The cognitive psychology perspective. In D. Holdzkom and P. B. Lutz, (eds.), *Research within reach: Science education* (pp. 171–189). Charleston, W.V.: Research and Development Interpretation Service, Appalachia Educational Laboratory.

Dewey, J. (1938). *Experience and education*. New York, N.Y.: Macmillan, Inc.

Feynman, R. P. (1985). *Surely you're joking, Mr. Feynman*. New York, N.Y.: Bantam.

Leyden, M. B. (1984). You graduate more criminals than scientists. *The Science Teacher, 51*(3), 27–30.

National Assessment of Educational Progress. (1978). *The third assessment of science, 1976-77*. Denver, Colo.: Author.

Pittendrigh, C. S., and Tiffany, L. H. (1957). *Life: An introduction to biology*. New York, N.Y.: Harcourt-Brace Jovanovich.

Simpson, G. G. (1963). Biology and the nature of science. *Science, 139*(3550), 81–88.

Weiss, I. R. (1987). *Report of the 1985–86 national survey of science and mathematics education.* Research Triangle Park, N.C.: Center for Educational Research and Evaluation, Research Triangle Institute.

Yager, R. E. (1985). The attitudes of the public toward science and science education. *Iowa Science Teachers Journal, 22*(2), 8–13.

CHAPTER 3

IS SCIENCE SINKING IN A SEA OF KNOWLEDGE? A THEORY OF CONCEPTUAL DRIFT

John Lochhead
Robert E. Yager

FIRST RUMBLINGS

Twenty-five years ago the ground on which we stood seemed solid and stable. Then suddenly it began to move. While evidence had been amassing for nearly 100 years, it was only when deep sea drilling data were obtained around 1970 that the majority of scientists realized continents were adrift (Sullivan, 1991). A cherished sense of security and permanence was lost forever.

Today a similarly unsettling realization has begun to spread throughout science education (Resnick, 1983). It is as if the most basic stepping stone of our profession, "knowledge itself," has shifted beneath our feet (von Glasersfeld, 1987).

This epistemological equivalent of plate tectonics is the theory called constructivism. It explains that all scientific and rational knowledge is a construct of the individual who claims to know it, and thus knowledge cannot be absolute in validity, or in strict sense, commonly held by the community of experts (Maturana, 1978). As with continental drift, this theory, which at first seems terrifying, turns out to be largely irrelevant to much of our daily lives. At the same time, it is extremely powerful in special applications. We must live as though the ground is permanent: most of the time we must act as though our knowledge is an accurate picture of the world. Yet understanding the manner in which continents move is critical to the work of earth scientists. Likewise, science teachers need to understand the dynamics of scientific knowledge: what it is, how it is formed, and how it changes (Mestre, 1994).

The nature of "knowledge" as we understand it today is substantially different from what it was only a decade or two ago (Feyerabend, 1978; Kuhn,

1970). The origins of this transformation of our perception of concept go back nearly 300 years, to 1710 and the Italian philosopher Vico (1858). Some roots of this view extend back even further to the pre-Socratic Greeks (Pirzig, 1984). As with the theory of continental drift, constructivism stems from several diverse disciplines of inquiry. These include: philosophy (Feyerabend, 1978, 1987; Lakatos, 1976), cybernetics (Glasser, 1986; Powers, 1973; von Glasersfeld, 1979*a*), language analysis (Maturana and Varela, 1988; Vygotsky, 1962), and developmental and cognitive psychology (Piaget, 1965; von Glasersfeld, 1979*b*).

Some Aftershocks

An important consequence of the constructivist perspective is that speakers, teachers, and writers need to pay careful attention to their audience. We, the authors of this chapter, may have painted in your mind an alarming picture of concepts adrift in a sea of uncertainty anchored only with nebulous ties to mysterious creeds such as cybernetics, cognitive psychology, and something called epistemology. We have created the threat that cherished beliefs will be yanked away from underneath your feet. While in a sense this is all quite accurate, we wish to comfort you with the observation that both continents and concepts change shape very slowly. To a first approximation, nothing will be much different from what it was before. In particular, excellent science instruction will remain excellent. The promise that the new constructivist theory holds is not so much in overthrowing current truths as in helping to understand the causes of the cracks and faults that criss-cross the sphere of our current activity. As we come to understand those faults we will see new ones, but it is a gradual process. There are few earthquakes or tidal waves.

The Epicenter

Constructivism involves first and foremost a shift in focus from what the teacher does to what the learner interprets. Constructivism is not a theory of teaching; it is a theory of learning and of that which is learned. Constructivism cannot dictate how we should teach, rather it informs us of how we should search for evidence concerning what it is that we have taught. From the constructivist perspective, how well the teacher teaches is inseparable from how well the students learn. This was reported by Collingwood (1961): "Oh undergraduates come up the vulture [teacher] did beseech, and let us see if you can learn as well as we can teach" (p. 362).

What is it that we know about learners today that was not known fifty or even twenty years ago? Many science educators trace their interest in constructivism to the work of Piaget. Starting in about 1920, Piaget began a study of children's intellectual development that established many of the basic principles of constructivism. He showed that from their earliest moments children engage in categorizing their experiences with respect to what they already know. Babies distinguish things that are good to suck from those that are not. Six-year-olds can separate objects that float from those that sink or animals that fly from those that swim. The categories that children can create depend on their knowledge (the phenomena they have experienced and remember) *and* on the logical structures they have developed. Some children will be uncomfortable with birds that swim or fish that fly. They have not yet learned to handle overlapping categories or to separate a characteristic (such as flying) from the essence of the object.

Through the 1960s, many saw constructivism as pertaining only to children. In science education the primary impact was on the sequence of concepts presented in the curriculum. Since children do not have all the logical structures of adults, it was considered important to adjust what children were told in order to keep it consistent with their current level of logical processing. To aid teachers in this endeavor, various tests were created to help teachers define the level of intellectual development attained by each child in his or her classroom (Karplus, Karplus, and Formisano, 1979; Renner, 1977). Unfortunately, the more carefully teachers looked at their students' reasoning, the less useful they found the tests. Student performance was found to be more varied and complex than the simplest interpretations of the tests suggested.

Piaget's studies, and most other research inspired by this work, examined what children knew *prior* to formal instruction. From the perspective of science education these studies provided an insight into what children already knew and into what they might find difficult to learn. It was hoped that a carefully designed science curriculum could use the building blocks that research indicated children already possessed and build from these the new concepts needed in science. This version of constructivism held that children could only learn new ideas by piecing together existing concepts, but it left to teachers much of the work of assembling the pieces.

Then in the late 1970s an ominous series of experiments began to undermine the granite foundations of science teaching. (Relax, this tremor will soon be over.) The techniques pioneered by Piaget were applied to students *after* they had completed a course of formal instruction. To nearly everyone's surprise the concepts held by students after instruction differed little from those they held prior to instruction (Clement, 1982; McCloskey, 1983). At the conceptual level the net impact of two semesters of college physics was shown to be arbitrarily close to zero. At first this was attributed to bad teaching. But experiments carried out in universities all over the world showed remarkably constant results across a wide range of teaching strategies and educational systems (Mestre and Lochhead, 1983). The way the teacher taught had very little impact on what the students learned (Hallerin and Hestenes, 1985a, 1985b).

Many proposed that the students had been misled by textbooks and instructors that were insufficiently precise. This weak instruction, presumably occurring in the early years (since college faculty would be incapable of such lax behavior), had poisoned the students' minds, making it difficult to teach them. But evidence continued to come in indicating that students were constructing ideas that could not possibly ever have been taught to them (Clement, Lochhead, and Monk, 1981). Students were constructing knowledge on their own, independent from the instruction they received. The situation appeared out of control.

The Origins of Conceptual Drift

It may be useful to pause here to distinguish, in yet one more way, constructivism from other educational philosophies. There are many approaches to education that, on the one hand, hold that it is better for students to learn by discovery or by concrete experience. Constructivists, on the other hand, are not necessarily pleased that students construct knowledge. Like everyone else, constructivists are often excited by what they know and would like to tell students about this knowledge. Yet they realize this can be a waste of time if students are not prepared to receive the message. Constructivism is not a utopian solution to education; it is more of a nuisance or curse. Students construct ideas whether we would like them to or not, and they do so without much regard for our input or feelings about the matter. We cannot stop the continents from drifting and we cannot stop students from constructing their own knowledge. It may not be convenient, but we must adapt to a world in which students have minds of their own.

Conception Currents

Constructivists have a simple interpretation of the data illustrating student misconceptions that are not altered by formal instruction. People can only interpret what they see, hear, or otherwise experience, in terms of what they already know and believe. If the authors of this piece were to thank the Martians for their assistance with the early drafts of this paper, most of you would assume either that we were crazy or that we were not referring to aliens from Mars. You have constructed a mental framework that defines how you interpret what we say to you. Students enter courses with similar mental frameworks, some much more rigid than your resistance to the idea of visiting space aliens who spend their weekends helping academicians write papers. For these students, sections of your lecture might just as well have been delivered in a Martian dialect.

Yet, not all lecturing is ludicrous; it is important to know when and when not to lecture. You need to understand your audience and the beliefs of your listeners. Some of their ideas are as firm as continental divides. Other ideas are very tentative. The impact of direct instruction depends on the extent to which it fits comfortably with what is already known. When we are told things that do not fit well with our current beliefs, we tend to alter what we hear until we get a good fit. Since this is often done subconsciously, we usually are not aware that we have altered the message from what the speaker had intended. Constructivists criticize lectures because they often provide speaker and audience with very different experiences without either party being aware of the difference. Lectures are an excellent way to help students pull together disjointed pieces of knowledge that they already comfortably understand; but they tend to be a poor way to introduce difficult new concepts. The same, unfortunately, may be said of written materials, particularly chapters like this one that attempt to introduce radically new ways of thinking.

In order to enhance conceptual growth it is critical for teachers to know a great deal about how the students interpret their experiences with new concepts (Baird and Northfield, 1992). It is here that constructivism and cognitive research have much to offer science instruction (Mestre, 1987). Constructivism provides the framework in which teachers can appreciate the need to be sensitive to the wide range of different ways in which students may interpret a lesson. Cognitive research can provide specific examples of the kinds of constructions students are likely to create as well as the kinds of concepts they may have formed prior to instruction (Helm and Novak, 1983; Novak, 1987, 1993).

The Core of Cognitive Theory

Let us assume that we want to teach students about making ice cream the old way. We submerge a container of cream in a tub of water and ice; then we add salt. It is important to know that most students already will have constructed theories of heat and cold, and that in these theories heat and cold act like liquids that flow from one location to another. Adding salt does not appear to be the equivalent of making a pathway for the cold to flow in, or the heat to flow out. For this reason the students are likely to be very puzzled. They will want to verify that the ice water really does get colder. An explanation based on describing how energy is absorbed during the melting process will be viewed skeptically. It does not fit the liquid flow model, nor does it help students see why they should give up on a theory that has worked well for them (as it has for most of us) for years and years.

To make progress beyond this point, it is first necessary to get the students to reconsider their concepts of heat and cold. They may need to encounter several situations in which the cold flow model has difficulty before they will feel it is worth the trouble to consider any alternatives. Fortunately teachers can read about the theories students are likely to have (Pfundt and Duit, 1991) as well as about some of the specific paths through which students might come to revise their theories of heat (Rogan, 1988). In many areas of science little or no work has been done to study how students imagine the key concepts. Here the teacher's best guide is to examine the history of how scientists themselves developed the concepts in the first place (Hills, 1992).

Attempts to Rejoin Divided Concepts

The idea of linking science, technology, and society was first proposed as a means of making science more interesting and of ensuring that it would be learned in a context of useful application. When these three elements are combined with history, a more fundamental reason for the linkage emerges. The best available guide for teaching new concepts is often their historical evolution, including the combination of their scientific, technological, and social applications. Studies of the manner in which students construct scientific concepts often show striking parallels to the historical record. There are, of course, significant differences, and each student will construct concepts in his or her own unique way. Yet, the differences are not so great as to make the experience of the past irrelevant.

The Social Implications of Concept Construction

The central idea of constructivism is that students cannot really learn by passively absorbing or copying the understandings of others. Instead all learners *must* construct their own understandings. Such understandings are organized by and related to existing knowledge—most of which was formed individually by each person based on his or her past experiences. Old concepts can only be displaced when learners engage in problem situations where their own constructed meanings are not adequate (Clement, 1987; Lochhead and Mestre, 1988; Minstrell, 1987). Social interaction in the form of discussions, debates, and arguments play a crucial role in challenging the adequacy of old concepts.

Our experience is strongly influenced by our social interactions with other people (as well as with animals, machines, and institutions). This is true even when we appear to think and act in a solitary manner. Most thought involves language, and language carries the memory of the social interactions through which it was constructed. Thus, while the ideas we construct are uniquely our own, the process through which they are formed may employ elements extracted from what other people have constructed before us.

Can We Harness the Energy of Concept Construction?

Effective teachers constantly encourage learners to encounter new situations that will challenge previously constructed explanations. Such disturbances provide situations that cause discomfort and cause further thought and perhaps new constructions. In these instances the teacher's role is to *confuse* so as to provoke reflection; it is, therefore, important to resist the temptation to *defuse* the tension and provide an answer.

Table 3.1 provides an outline of some of the activities a teacher can use to promote learning from within the constructivist paradigm. However, the most critical ingredient is not the list of activities but the style in which each is employed. From the constructivist perspective it is not reasonable to view oneself as a fountain of information with moral responsibility for transmitting that information to students. The activities in Table 3.1 should be conducted by the students themselves. Teachers should be mostly observers, moderators, and instigators.

The constructivist perspective suggests that it would not be helpful for us to outline the specific techniques and procedures of an effective teacher. Teachers must deal with individual students and get them engaged in meaningful activities. This involves attitudes more than techniques or procedures. It is, perhaps, most useful to characterize instruction as a focus on the five *E*'s: engagement, exploration, explanation, elaboration, and evaluation. For each *E* it is the student who must act—not *just the teacher.*

Specific activities that could be useful for constructivist learning include:

1. encouraging and accepting student autonomy, initiation, and leadership (Glasser, 1986).
2. allowing student thinking to drive lessons. Shifting to content and instructional strategies that are based on student responses.[1]
3. asking students to elaborate on their responses (Doris, 1991).
4. allowing wait time after asking questions (Rowe, 1973).
5. encouraging students to interact with each other and with you (Whimbey and Lochhead, 1991).
6. asking thoughtful, open-ended questions.
7. asking students to articulate their theories about concepts before accepting teacher (or textbook) explanations of the concepts (Clement, 1987; Minstrell, 1987).

8. looking for alternative concepts of students, and designing lessons to address any misconceptions (Yager, 1991).

None of the above processes is new to teaching, but the constructivist perspective gives them new meaning and importance. Originally they were favored as means of providing students with greater autonomy and responsibility, rea-

TABLE 3.1 Constructivist Strategies for Teaching

Invitation:	Observe one's surroundings for points of curiosity Ask questions Consider possible responses to questions Note unexpected phenomena Identify situations where students perceptions vary
Exploration:	Engage in focused play Brainstorm possible alternatives Look for information Experiment with materials Observe specific phenomena Design a model Collect and organize data Employ problem-solving strategies Select appropriate resources Discuss solutions with others Design and conduct experiments Evaluate choices Engage in debate Identify risks and consequences Define parameters of an investigation Analyze data
Proposing Explanations and Solutions:	Communicate information and ideas Construct and explain a model Construct a new explanation Review and critique solutions Utilize peer evaluation Assemble multiple answers/solutions Determine appropriate closure Integrate a solution with existing knowledge and experiences
Taking Action:	Make decisions Apply knowledge and skills Transfer knowledge and skills Share information and ideas Ask new questions Develop products and promote ideas Use models and ideas to illicit discussion and acceptance by others

sons that were largely political or moral. Constructivism, however, takes a more pragmatic stance. Even if your goal were to be as coercive as possible, these would be the most effective strategies for cramming real science into your students' heads. Unless you compel students to critique their own constructions, through mechanisms such as discussion with classmates, the concepts you want to convey have virtually no chance of getting farther than the tip of your tongue. Without the simplistic illusions associated with the knowledge transmission model, concept development becomes a complex process. In Table 3.2 we highlight some of the progression of stages learners need to pass through on their way to understanding scientific concepts. It starts with a self-analysis and moves to increasingly more inclusive levels of consensus. In practice the progression may look more like a random hopping back and forth between levels, but the overall effect should be integrated.

Analyzing Sedimentary Textbooks

Constructivism has major implications for curriculum frameworks and textbooks. Most existing courses and textbooks are organized around concepts that someone else has decided students should master.[2] They are not designed to encourage an open-ended investigation of topics that interest the students. In a con-

TABLE 3.2 Constructivist/STS Grid

	Who	*Problems*	*Responses*	*Results*
A.	Individual student	Identifying problem	Suggesting response	Self-analysis
B.	Pairs of students	1. Comparison of ideas 2. Resulting questions	Agreeing on approach to problem(s)	Two-person agreement
C.	Small group review	1. Consider different interpretations 2. Achieve consensus	1. Consider different responses 2. Achieve consensus	Small group consensus
D.	Whole class (local community)	1. Discussion 2. Identify varying views	Acts to gain consensus	Whole class agreement
E.	Science community	Comparison of class views vs. those of scientists	Comparison of class views vs. those of scientists	Consensus/new problems/actions

structivist classroom the function of textbooks and other support materials would be very different from most current practices. Instead of defining what should be learned, to what depth, and in what order, textbooks ought to provide a source of data that students would be expected to probe rather than ingest. Ideally, such books might include a diversity of perspectives, even conflicting opinions. They should not strive to be complete but rather leave to students the task of deciding how and where to search for additional data. Students should also confront the task of evaluating sources; for this reason textbooks ought to contain errors and information of dubious quality.[3] Fortunately, most textbooks contain ample amounts of dubious information even without the benefit of constructivism.

Textbooks should ask questions, pose problems, suggest areas for inquiry and actions to advance the inquiry. Textbooks should help students—just as teachers should. But this help should not involve simply telling the reader what to believe or what facts to memorize for the exam. The Structured Inquiry Approach (College Board, 1990) is one example of constructivist instruction that provides assistance and support while encouraging independent thought.

Measurement and Prediction of Conceptual Tremors

Assessment is a constant challenge to constructivist teachers. While remembering facts and procedures is relevant to learning, recall is only a small part of mastery and often reflects no more than a superficial engagement with the content. Measurement of the five *E*'s (engagement, exploration, explanation, elaboration, and evaluation) provides a more meaningful assessment of learning, but here too the results can be misleading (Perry, 1970). To the constructivist, assessment is always uncertain. Every measurement should be subjected to scrutiny. Only after a range of different kinds of assessment reveal a pattern can tentative stock be put in the validity of the results. This note of caution creates a serious dilemma for most teachers. In reality, grades must be assigned from a limited data set, far less than is adequate. But because constructivism indicates that *all* decisions are based on uncertain knowledge, it provides no excuse for inaction. The constructivist teacher strives to treat with dignity and respect those students who must be failed, and remains constantly aware that some of the A's should have been F's and some of the F's should have been A's.

The Wave of the Future

The biggest paradigm shift required by constructivism concerns new roles for teachers and new demands placed on teacher education. Teachers must learn to master a broader set of roles, behaviors, and strategies. The best teachers are themselves active learners; they always strive to better understand their own teaching and the effect it has on student learning. This is a continuing process; a type of learning that is never complete and never dull. Constructivism teaching always provides an exciting adventure; courses where traditional teaching is perceived as routine or unrewarding can be transformed into stimulating challenges (Lochhead, 1991).

Most science teachers liked the science they experienced in school and proceeded to college majoring in science because of this liking and the success they experienced with their subject. Science teaching in secondary school is modeled after the teaching teachers experienced during their own college preparation. But college faculty rarely employ constructivist instructional practices. Professors profess; they give lectures on topics *they* view as important. They expect students "to learn" the material presented in lecture and assess student learning largely on the basis of factual recall. Introductory laboratories emphasize following instructions and finding results, where the desired result is known well before the "investigation" begins. While university-based scientists often blame education programs for what they believe are poorly trained teachers, most prospective science teachers take far more courses in science than in education. Breaking out of the old mold will require either a change in the manner in which university science is taught or a reduction in the degree to which prospective teachers take science at the university.

NOTES

1. The Classtalk project (Abrahamson, 1991) demonstrates how technology can be used to maintain such responsiveness even in large lecture situations. A low technology alternative is described by Light (1990).

2. An encouraging departure is the text created by the Hawaii Algebra Learning Project. This book is the result of a ten-year study of student concepts and a careful testing procedure that analyzed a range of student thinking in every area of the algebra curriculum.

3. Excellent examples of how this approach can be employed to good effect are found in some texts produced by the Peoples' Education Movement in South Africa. *The Right to Learn* (Christie, 1985) contained "conservative," "moderate," and "radical" viewpoints on many of the issues discussed. While the author undoubtedly believed that some of these perspectives had less than dubious origins, she respected her readers' capabilities to interpret the data for themselves.

REFERENCES

Abrahamson, L. A. (1991). *Pedagogical techniques for classroom communications systems*. (Final Project Report for NSF award ISI-9060464). Yorktown, Va.: Better Education Inc.

Baird, J. R., and Northfield, J. R. (eds.). (1992). *Learning from the PEEL experience.* Melbourne: Monash University Printing.

Christie, P. (1985). *The right to learn.* Johannesburg, South Africa: Ravan Press.

Clement, J. (1982). Students' preconceptions in introductory mechanics. *American Journal of Physics, 50,* 66–71.

Clement, J. (1987). Overcoming student's misconceptions in physics: The role of anchoring intuitions and analogical validity. In J. D. Novak (ed.), *Proceedings of the second international seminar on misconceptions and educational strategies in science and mathematics,* Vol. 3 (pp. 84–97). Ithaca, N.Y.: Cornell University, Department of Education.

Clement, J., Lochhead, J., and Monk, G. S. (1981). Translation difficulties in learning mathematics. *American Mathematical Monthly, 88*(4), 286–290.

College Board. (1990). *Academic preparation in science* (2d ed.). New York: Author.

Collingwood, S. D. (ed.). (1961). *Diversions and digressions of Lewis Carroll.* Dover, N.Y.: Dover.

Doris, E. (1991). *Doing what scientists do children learn to investigate their world.* Portsmouth, N.H.: Heinemann Education.

Feyerabend, P. (1978). *Against method.* Norfolk, Great Britain: Cambridge University Press.

Feyerabend, P. (1987). *Farewell to reason.* New York: Verso.

Glasser, W. (1986). *Control theory in the classroom.* New York: Harper and Row.

Hallerin, I., and Hestenes, D. (1985a). Common sense concepts about motion. *American Journal of Physics, 55,* 1056–1065.

Hallerin, I., and Hestenes, D. (1985b). The initial knowledge state of college physics students. *American Journal of Physics, 55,* 1043–1055.

Helm, H., and Novak, J. D. (1983). *Proceedings of the international seminar on misconceptions in science and mathematics.* Ithaca, N.Y.: Cornell University.

Hills, S. (ed.). (1992). *The history and philosophy of science in science education, Volumes I and II: Proceedings of the second international conference.* Kingston, Ontario: Queens University.

Karplus, R., Karplus, E., and Formisano, M. (1979). Proportional reasoning and the control of variables in seven countries. In J. Lochhead and J. Clement (eds.), *Cognitive process instruction* (pp. 47–103). Hillsdale, N.J.: Lawrence Erlbaum Publishers.

Kuhn, T. (1970). *The structure of scientific revolutions.* Chicago: University of Chicago Press.

Lakatos, I. (1976). *Proofs and refutations: The logic of mathematical discovery.* Norfolk, Great Britain: Cambridge University Press.

Light, R. I. (1990). *The Harvard assessment seminars, first report.* Cambridge: Harvard University (HGSE).

Lochhead, J. (1991). Making math mean. In E. von Glasersfeld (ed.), *Radical constructivism in mathematics education* (pp. 75–87). The Netherlands: Kluwer Publishers.

Lochhead, J., and Mestre, J. P. (1988). From words to algebra: Mending misconceptions. In A. Coxford and A. Schulte (eds.), *The idea of algebra K–12: National Council of Teachers in Mathematics education year book, December* (pp. 127–135). Reston, Va.: National Council of Teachers in Mathematics.

Maturana, H. R. (1978). Biology of language: The epistemology of reality. In G. Miller and E. Lenneberg (eds.), *Psychology and biology of language and thought* (pp. 27–63). New York: Academic Press.

Maturana, H. R., and Varela, F. J. (1988). *The tree of knowledge: The biological root of human understanding.* Translated by R. Baldick. Boston: New Science Library Shambala Press.

McClosky, M. (1983). Intuitive physics. *Scientific American, 248,* 122.

Mestre, J. P. (1987). Why should mathematics and science teachers be interested in cognitive research findings? *Academic Connections, Summer,* 3–5, 8–11.

Mestre, J. P. (1994). Cognitive aspects of learning and teaching science. In F. J. Fitzsimmons and L. C. Kerpelman (eds.), *Teacher enhancement for elementary and secondary science and mathematics: Status, issues, and problems* (pp. 3-1–3-53) (Report # NSF 94-80). Washington, D.C.: National Science Foundation.

Mestre, J. P., and Lochhead, J. (1983). The variable-reversal error among five cultural groups. In J. Bergeron & N. Herseovics (eds.), *Proceedings of the fifth annual meeting of the North American chapter on the International Group of the Psychology of Mathematics Education* (pp. 180–188). Montreal, Quebec: Group on Psychology of Mathematics Education, North American Chapter.

Minstrell, J. (1987). *Classroom dialogues for promoting physics understanding.* Paper presented at the Annual Meeting of the American Education Research Association. Washington, D.C.

Novak, J. D. (ed.). (1987). *Proceedings of the second international seminar on misconceptions in science and mathematics.* Ithaca, N.Y.: Cornell University.

Novak, J. D. (ed.). (1993). *Proceedings of the second international seminar on misconceptions and educational strategies in science and mathematics.* Ithaca, N.Y.: Cornell University.

Perry, W. G. (1970). *Forms of intellectual and ethical development in the college years: A scheme*. New York, N.Y.: Holt Rinehart and Winston.

Pfundt, H., and Duit, R. (eds.). (1991). *Bibliography: Students' alternative frameworks and science education*. Germany: University of Kiel.

Piaget, J. (1965). *The child's conception of number*. New York: W. W. Norton and Company.

Pirzig, R. M. (1984). *Zen and the art of motorcycle maintenance*. New York: Bantam.

Powers, W. T. (1973). *Behavior: The control of perception*. Hawthorne, N.Y.: Aldine Publishing Company.

Renner, J. W. (1977). *Evaluating intellectual development using written responses to selected science problems*. Unpublished report to the National Science Foundation (EPP75-19596).

Resnick, L. B. (1983). Mathematics and science learning? A new conception. *Science, 220*, 477–478.

Rogan, J. M. (1988). Development of a conceptual framework of heat. *Science Education, 71*(1), 103–113.

Rowe, M. B. (1973). *Teaching science ad continuous inquiry*. New York: McGraw-Hill.

Sullivan, W. (1991). *Continents in motion* (2d ed.). New York: American Institute of Physics.

Vico, G. (1858). *De antiquissima Italorum sapientia*. (original work published 1710). Naples, Italy: Stamperia de' Classici Latini.

von Glasersfeld, E. (1979*a*). Cybernetics, experience and the concept of self. In M. N. Ozer (ed.), *A cybernetic approach to the assessment of children: Toward a more humane use of human beings* (pp. 61–110). Boulder, Colo.: Westview Press.

von Glasersfeld, E. (1979*b*). Radical constructivism and Piaget's concept of knowledge. In F. B. Murray (ed.), *The impact of Piagetian theory on education, philosophy, psychiatry, and psychology* (pp. 109–122). Baltimore: University Park Press.

von Glasersfeld, E. (1987). *The construction of knowledge*. Seaside, Calif.: The Systems Inquiry Series, Intersystems Publication.

Vygotsky, L. S. (1962). *Thought and language*. (E. Haufmann and G. Vakar, trans.). Cambridge: MIT Press.

Whimbey, A., and Lochhead, J. (1991). *Problem solving and comprehension* (5th ed.). Hillsdale, N.Y.: Lawarence Earlbaum Associates.

Yager, R. E. (1991). The constructivist learning model: Towards real reform in science education. *The Science Teacher, 58*(6), 52–57.

CHAPTER 4

THE CONGRUENCY OF THE STS APPROACH AND CONSTRUCTIVISM

Martha Lutz

WHAT IS LEARNING?

In his book *Learning and Instinct in Animals*, Thorpe (1963) defines learning as "that process which manifests itself by adaptive changes in individual behavior as a result of experience." This definition has two points that bear emphasis in the context of this chapter. One point is the definition of learning as a process. This implies that learning is active—and so it should be: the learner should be mentally active. Passive reception of information is *not* learning. The second point is that learning is the result of experience. This implies that the learner must have experiences in order to learn. If we accept Thorpe's definition, then any learning event must include opportunities for the learner to be mentally active and to have concrete experiences.

Traditional teaching, in which the teacher discourses at the students while the students passively receive information, does not fit the definition above. Is there any teaching strategy that would allow for active involvement on the part of the students, as well as a chance to gain the experiences that Thorpe's definition calls for? Are there other definitions of learning that shed more light on appropriate teaching strategies?

LEARNING THEORY AND TEACHING PRACTICE

Driver and Oldham (1986) view learning as a change in conceptions. This change in the way knowledge is structured by the individual learner is known as construction of knowledge. The formal name for this view of learning is constructivism.

Constructivism is a learning theory. But a theory requires a mechanism. What is the mechanism of constructivism? For the student, the mechanism is conceptual change, which takes place within the mind of the individual learner. For the teacher, one viable mechanism is a teaching strategy known as STS. Briefly defined for this discussion, STS is the learning of science concepts in the context of real life experiences and with application to real life problems and issues.

This is not a new concept: in the 1950s, Benjamin Bloom asserted that knowledge that is organized and interrelated is better learned and retained than knowledge that is specific and isolated (Bloom, 1956). Traditional textbook driven teaching tends to be specific and isolated; STS teaching is by definition organized and interrelated.

Constructivism is the learning theory that underlies the teaching strategy known as STS. Viewed this way, constructivist teaching methods are by definition congruent with STS teaching methods.

The Student: Constructivism Requires Responsibility

One key feature of constructivism is that it requires learners to take responsibility for their own learning. Unlike traditional teaching approaches, in which the teacher is an omnipotent fount of information while the student is a passive recipient, effective constructivist teaching makes demands on the learner. If the learner is to actively construct knowledge, the learner must be actively involved in the learning process.

This sounds excellent. However, this optimistic scenario ignores the reality familiar to anyone who has spent much time in typical grade-school classrooms. Our students have been trained and suppressed from their earliest experiences in school, and the only role they are familiar with is the passive role. We tell them to sit still. We tell them not to talk in class, unless requested to regurgitate some specified lump of information. This is the format our students know well, and although they do not love it, they are comfortable with it. Anything unfamiliar is also uncomfortable; any proposed change will be met with resistance. But if the change is worth making, it is worth our while to ride out and overcome the resistance.

Our students are accustomed to being induced to learn. They are accustomed to having certain concepts presented at a certain pace and with certain almost ritualized quizzes and exams administered at predictable times. It is, if you will, a kind of intellectual force-feeding. There is no time and no permission for exploration, or for following natural curiosity; there is no provision for spontaneous learning, which is the most intriguing form of learning. While emphasizing quantity of content, we have lost our sense of the importance of quality.

When our students fail to take responsibility for their own learning, is this a failure on their part? Or is it a failure of the system, the style of teaching most commonly in use? Hawkins (1983) declares that it is a failure of the system. He insists, gently but firmly, that mathematics and science can be built on experiences that are most common in the background of all schoolchildren, regardless of social and economic environment.

Not all children have read *Peter Pan*. Not all have seen *Fantasia*, or listened to *Peter and the Wolf*. Some have no personal experience with geography or with history. Religion is an important but highly variable factor in the background of the individual children who face our teachers daily across the desks in the nation's schools. But virtually all these children—and the teachers too—have seen the sun rise and felt its warmth on the back of their necks. Most have watched snow fall, "sticking" on grass and other vegetation while melting on sidewalks and other hard surfaces such as blacktop. They have heard different birds sing different songs; heard, but perhaps paid no heed—wrapped up in the dry details of the Hardy-Weinberg equation or some other isolated bit of science content.

Given that this is true, we could take advantage of the common experiences of all human beings, including children, to allow teachers and their young students to explore, to indulge in spontaneous rather than induced learning. Then, perhaps, students will demonstrate that they can and will take responsibility for their own learning.

It is worth the risk. The traditional textbook-lecture system of induced learning has been a failure. Poor retention and inability to apply learned information are characteristic of current educational practices (Yager, 1989-1990). The Chautauqua program in Iowa is yielding exciting data that indicates that STS teaching improves both retention of and ability to apply knowledge, and engenders more positive attitudes toward science.

A Teaching Strategy for the Constructivist Approach

What teaching strategy, if any, will encourage students to take responsibility for their own learning? Is there any form of teaching that is in essence dynamic rather than static, and that inherently permits the flexibility of classroom activities necessary if students are to construct knowledge?

Such a teaching strategy does exist. It is a strategy that places emphasis on the process of arriving at an answer, rather than simply requiring that students be able to regurgitate the "right" answer—whether or not they understand either the answer or its justification. It is a strategy that allows students time to construct knowledge, thus circumventing the pitfall succinctly described by Hawkins (1983): "*In*struction by a teacher fails without a matching *con*struction by the learner."

STS teaching provides an appropriate strategy. An STS teaching strategy focuses on process, not product. This is consistent with constructivist thinking:

in order to construct knowledge, the learner must sort and organize new data. This can only happen if the process of generating that data is explicit. Otherwise, students may at most retain the "facts" given to them by the textbook or instructor, and not connect this information with real-life situations or known phenomena. According to Yager (1991): "Most persons have misconceptions about nature; typical schooling is ineffective in altering misconceptions; many of the most able students have as many misconceptions about science as the average high school student" and "students who score well on standardized tests are often unable to successfully integrate or contrast memorized facts and formulas with the experience-based interpretations they have acquired prior to instruction" (p. 53). Typical schooling is ineffective; early results of studies on STS classrooms indicate that STS teaching is effective.

Constructivism: Connecting With, Not Building On

Constructivism and conceptual change are complementary. For conceptual change to occur, the individual learner must be actively and responsibly involved in assembling new and old data in a way that makes sense. The new picture that emerges from this process of creative restructuring is the product of conceptual change: constructed meaning. Constructivism is often referred to as "building on" prior knowledge. It is—in a sense—but only in the sense that prior knowledge is *connected* to new data during the process of creative construction of meaning. It does not mean that in order for a learner to learn certain concepts, that learner must first master certain specified prerequisites and build *on* those in the manner of building a house on top of a foundation.

Perhaps the fault lies with the chosen semantics: "construct" sounds too much like a synonym for "build." There is a great difference, however, between building by accretion on a foundation that has already been laid, and constructing a personal knowledge structure out of new incoming data and making multiple connections between this new knowledge and preexisting mental constructs.

The essence of constructivism is that new incoming data are assembled so that the individual learner personally creates a mental structure that incorporates and provides a context for both the new data and its relationships to data already stored in long-term memory. Thus one does not construct knowledge by "building on" prior learning via accretion. Rather, one constructs knowledge via an act of creative assembly and restructuring.

The critical difference is the difference between *building on* and *connecting with*; the former implies prerequisites; the latter makes no such demands. The former implies a critical or even magical sequence for curriculum; the latter allows flexibility, opportunism in determining curriculum, and time for exploration. The former is congruent with teaching from textbooks; the latter is congruent with STS teaching.

Constructivist philosophy is not an intrinsic quality of a particular sequence of activities. Rather, constructivist philosophy is an intrinsic quality of a teaching strategy that must be applied individually to each activity that the teachers and students share in the classroom. Constructivism is not dependent on having certain specified prerequisites. It is dependent on the individual learner, who must take responsibility for his or her own learning and—via an energetic, creative effort—structure a coherent picture out of new and old data.

Is STS Teaching Congruent with Constructivism?

Roth (1989) sets up a straw man of antagonism between STS and constructivism. She suggests that the two are mutually exclusive. Her grounds are untenable: addressing current issues is *not* automatically mutually exclusive with constructivism. Use of relevant context for learning can certainly be constructivist. One could argue very persuasively that the most likely cause of conceptual change is when current issues are under investigation, and result in a perceived need for a more adequate explanation or viewpoint than that which the observer currently holds.

This perceived need is one of the necessary preconditions for conceptual change mentioned by Strike and Posner (1982). For example, students are more generally familiar with the concept of physical fitness than with that of Darwinian fitness. A common preconception about "survival of the fittest" is the assumption that "fittest" implies strongest, fastest, or biggest. An STS unit on adaptations of insects to new habitats could lead to a conceptual change to the understanding that "fittest" implies ability to reproduce successfully in a particular set of environmental conditions, and might have nothing at all to do with size, strength, speed, or other familiar criteria for being physically fit. (A classic example is the peppered moth in England during the Industrial Revolution, but every community would have numerous examples available concerning local species and environmental conditions.)

Roth (1989) makes unsupported assumptions about what an STS unit on photosynthesis might be like. On the basis of her own imagined hypothetical scenario, she condemns STS teaching and declares that it is both different from and inferior to conceptual change. This is an objectionable use of a hypothetical scenario. The end—a glorification of conceptual change—does not justify the means—a misleading condemnation of STS teaching. One valuable outcome of Roth's emphatically favorable analysis of constructivism is that it applies also to STS, in spite of her erroneous assertions that the two are antagonistic. Once that single false premise is recognized and removed, all her arguments in favor of constructivism also weigh in favor of the STS teaching strategy as advocated by Yager (1991).

Her description of how she taught photosynthesis to her own fifth-grade students exactly matches the general pattern for an STS teaching unit. Her assertion that this teaching method produces students who are able to "make predictions and observations, to change and develop explanations, and to apply ideas in a meaningful time frame" is an excellent testimonial of the benefits of STS teaching.

CHARACTERISTICS OF STS AND CONSTRUCTIVISM

What characterizes STS teaching? An STS approach starts with students' questions and interests, rather than with the magical next lesson in a predetermined sequence. Once the students have identified an area of interest, the teacher—in the role of a facilitator, rather than a fount of knowledge—helps the students to refine a specific question or topic for investigation. The students determine what resources to use, whether or not to conduct experiments, and proceed to gather information that will help them address the question or issue they have selected.

All of the information they collect is in the context of the issue they are investigating. This use of context is a key characteristic of STS (Yager, 1991). The students are learning science in the context of real experiences. This includes the technological applications of science. Not only do the students have a context for the data they collect, they are also required to analyze, synthesize, and evaluate that data. These are all higher order thinking skills, and are also steps characteristic of the process of constructing knowledge. Knowledge thus obtained will be understood and retained better than facts memorized and regurgitated purely for the purpose of passing a test in science class (Reinsmith, 1993). Knowledge constructed via an STS approach will be available to the learners as they attempt to cope with real life situations, not just classroom tests.

Learners of all ages and in many diverse situations are capable of retaining unconnected bits of information, but are generally unable to apply this knowledge (Roth, 1989). A student may "learn" all about the reactions that take place in photosynthesis, and yet not realize that plants capture energy from the sun and convert it into chemical energy, which animals can then acquire (by eating) and use to do things like run, find mates, and build homes. Plants are directly solar-powered; animals are indirectly solar-powered. Knowing the chemical reactions of photosynthesis cannot guarantee that an individual learner will be able to apply that knowledge or synthesize it with other knowledge.

Inability to apply or manipulate knowledge is characteristic of information acquired without a context. Reinsmith (1993) asserts that "Learning cannot take place outside an appropriate context." He emphatically states that there is no such thing as learning in isolation, and yet all too often that is exactly what we attempt to do: create learning in the complete absence of any context outside the textbook. This violates another of Reinsmith's ten fundamental truths: "Real learning connotes *use.*"

Context is critical to learning. How often do we provide a context for the knowledge that we attempt to impart to our science students? And how often do we provide students with opportunities to apply their newly constructed knowledge?

STS teaching by definition provides a context for the information that students acquire. It is the context that drives the search for relevant information: learning is context-driven, rather than textbook-driven. This is a characteristic of a constructivist approach: allowing student's questions and preconceptions to drive the lessons. STS also addresses student's alternative concepts, including misconceptions—another characteristic of a constructivist approach.

For instructors eager to use a constructivist approach, STS provides a teaching strategy that is both effective and flexible. The six basic steps are:

1. Brainstorm an issue or topic;
2. Define a specific question or phenomenon;
3. Brainstorm resources for obtaining information;
4. Use the resources to collect information;
5. Analyze, synthesize, evaluate, create; and
6. Take action.

This is not to imply that these steps must be taken in magical sequence, from one to six. Any step subsequent to step one can feed back into a repetition of any earlier step. Each individual investigation will have its own unique sequence and combination, and variations on the themes listed above.

Note that all of the above are steps taken by the students. The teacher is primarily a facilitator. When the students take action, their action should be based on substantiated information discovered and learned through their own studies and experiments. (They do not take action based on opinions unsupported by evidence.)

Another key characteristic of STS teaching compensates for another truth of learning that is well known but rarely taken into account by curriculum planners: the individual nature of perception. As Cleminson (1990) points out: "Despite the best intentions of curriculum planners and teachers, learning in science is a very personal activity and different students may well learn different things from the same learning experiences." A natural extension of this viewpoint is that an essential element of constructivist learning is the opportunity for students to share and negotiate their understanding of what they have experienced. The underlying premise is that group learning promotes appropriate conceptual change. Such group learning is an inherent feature of STS teaching. Both constructivism and STS teaching are characterized by group learning.

Does STS Really Work?

The only possible answer to this question must come from real experiences of STS teachers. Is it possible for students to learn science content via STS investigations? Will they really take responsibility for their own learning? Can they carry out the steps listed above with the teacher in the role of facilitator rather than lecturer? Will the students take meaningful action?

One representative answer comes from a school in Council Bluffs, Iowa (Lutz, 1992), where a ninth-grade chemistry class was being distracted by the particularly violent rainstorm going on outside the window. The teacher recognized the wisdom of allowing the students to watch the torrents, rather than trying to divert them forcibly—and perhaps unsuccessfully—to a discussion of careers in chemical engineering. The students and the teacher gathered, talking, at the rear of the room and watched the storm. One of the students asked about the acid content in rain. Taking advantage of the students' natural curiosity (a fine example of "the teachable moment"), the teacher used this question to launch an STS investigation about acid rain.

One student went out and collected a sample of rain to begin the activities. The students generated a list of potential directions for study:

1. How acidic is our rain?
2. Is the rain more acidic on the other side of the bluffs?
3. Do the industries in Omaha add pollutants to our rain?
4. Does the rain have more acid at the start of the storm or at the end?
5. Does snow or hail have acid?
6. If our rain is acidic, can we clean it up?
7. What is pH and how do we measure it?
8. Will acid rain hurt people? plant life?
9. Is rain in puddles on the ground the same pH as rain falling from the sky?

The class embarked on a flurry of activities, centered on research, surveys, readings from current literature, and lab activities. The students decided to test water from other areas, and wrote letters to other schools asking them to send rainwater samples. Soon many different resource people were involved: a speaker from the department of Game and Fish, a local waste management specialist, and experts in Washington, D.C., contacted by telephone. (At one point in their research, the students told the teacher to "Go sit in a corner and not bother them . . . they needed time to go through their reading material and make sense of some tough questions"!)

The students prepared pamphlets, synthesizing the results of their research. These pamphlets were designed to help family members, neighbors, city government people, and classmates understand acid rain. The five best pamphlets were reproduced and distributed at a city council meeting. As their final activity, the students wrote letters of thanks to all the speakers, resource people, and to the schools that had sent water samples.

The students learned science, and related what they learned to both societal issues and to their own lives. They were also able to have an impact on their community.

Is STS magic? No. Will it always work as well as it did in the example above? Probably not. But the story from Council Bluffs is only one of many, many success stories that come from Iowa. The story above illustrates that STS teaching has the potential to provide a mechanism for putting constructivist learning theory into practice. It does really work.

Why Nurture STS?

Even if you accept that STS really does work, that it encourages students to take responsibility, and that it is congruent with constructivism, is this sufficient reason to adopt STS teaching? There are surely other teaching strategies that would also be effective. It would be unlikely that any single teaching strategy would be the magical and only approach that leads to effective reform of science education in our country.

STS works; What other teaching styles might work? This is a reasonable and open-minded attitude, but it does not preclude using the STS approach. Because STS does work, we have a moral obligation to nurture it. The best argument for this is one made in general terms by Hawkins (1983) that is worth repeating at length: "In the biological world, there is a kind of theorem about the emergence of new types and new species: they do not arise by sudden or massive modification of older and dominant species, but appear, rather, because some genetic variety, initially rare, manages to grow at a rate faster than others around it when a new ecological niche appears that favors it. That is why evolution is not predictable; it works through the amplification of initially rare varieties that early go unnoticed. Good ideas have that kind of history, too, and so we often see it only in retrospect. I should hope that we begin to think of the improvement of education in just these terms. If good patterns of work with young children can be recognized early, while they are—as indeed they are—still rare and all too easily lost, we can hope to create for them some systems of protection, of support, and of growth" (pp. 83-84).

It is time for us to regard STS as a rare and valuable variety of teaching that deserves our protection and support. The question is not whether STS is the only effective teaching strategy, or whether any other style of teaching would be congruent with constructivism. It is enough for us to recognize that STS fulfills these criteria. And with this recognition, we acknowledge that our hope for more appropriate science education depends on nurturing STS, along with other rare, valuable, and effective teaching varieties that have the potential to grow and flourish.

Summary

A constructivist approach can be achieved via an STS teaching strategy; the result will be students who are independent learners, prepared to apply their knowledge to real-life situations, and who are capable of making informed decisions about their lives in a world of modern science and technology.

References

Bloom, B. S. (ed.). (1956). *Taxonomy of educational objectives (Handbook I: Cognitive domain)*, New York: David McKay.

Cleminson, A. (1990). Establishing an epistemological base for science teaching in the light of contemporary notions of the nature of science and of how children learn science. *Journal of Research in Science Teaching*, *27*(5), 429–445.

Driver, R., and Oldham, V. (1986). A constructivist approach to curriculum development in science. *Studies is Science Education*, *13*, 105–122.

Hawkins, D. (1983). Nature closely observed. *Daedalus*, *112*(2), 65–89.

Lutz, M. (1992). Carpe diem: Acid rain. *Scope, Sequence and Coordination Iowa News*, *3*(2), 10.

Reinsmith, W. A. (1993). Ten fundamental truths about learning. *The National Teaching and Learning Forum*, *2*(4), 7–8.

Roth, K. J. (1989). Science education: It's not enough to 'do' or 'relate.' *American Educator, Winter*, 16–48.

Strike, K. A., and Posner, G. J. (1982). Conceptual change and science teaching. *European Journal of Science Education*, *4*(3), 231–240.

Thorpe, W. H. (1963). *Learning and instinct in animals* (2d ed.). London: Methuen.

Yager, R. E. (1989–90). Appropriate science for all—an old goal unfurled anew. *Journal of the National Science Supervisors Association, 1*(1), 10–13.

Yager, R. E. (1991). The constructivist learning model: Towards real reform in science education. *The Science Teacher, 58*(6), 52–57.

Part II

What an STS Approach Can Accomplish

The six assessment domains in the Iowa Scope, Sequence, and Coordination Project (Iowa SS&C) have been used extensively in Iowa and a host of other places. The Iowa Chautauqua Model is being used across the United States as a way of initiating STS in kindergarten through twelfth-grade settings. The six assessment domains provide a map of how success in science classrooms can be measured. These domains include:

1. Concept domain (mastering basic content constructs);
2. Process domain (learning the skills scientists use in sciencing);
3. Application and connection domain (using concepts and processes in new situations);
4. Creativity domain (improving in quantity and quality of questions, possible explanations, and predicted consequences);
5. Attitude domain (developing more positive feelings concerning the usefulness of science, science study, science teachers, and science careers); and
6. World View domain (how the efforts assist students with an understanding of and ability to use basic science, that is, questioning, explaining, and testing objects and events in the natural world).

In this section a look at gender effects, concern for low- and high-ability students, and successes with other underrepresented groups in science provide added information concerning the power of STS. Although many of the extensive results come from Iowa where STS has been a focus for over ten years, supporting research is now beginning to appear from a great many other sources. Replication and use of the findings in this section are hastening moves to STS outside the United States as well as encouraging new research collaboration.

Chapter 5

Mastery of Basic Concepts

Lawrence Myers

Getting Started

Since the research efforts reported by Helgeson, Blosser, and Howe (1977), Stake and Easley (1978), and Weiss (1978), a spirit of reform in science education has swept across this nation and around the world. These three massive studies assessed the then present state of science teaching in schools across the nation. Science teachers, science educators, and school administrators were surveyed and their perceptions of school science were recorded. Research results were synthesized. Ethnographers studied what was occurring in representative science classrooms over a significant period of time. These studies helped to identify needed new directions for school science as a result of the deplorable situations found to exist. The reports of these three significant studies as well as a series of calls for specific reforms were reviewed and analyzed by Harms and Yager (1981) into a report for science educators. The Harms and Yager report of Project Synthesis confirmed the fact that science, as it was taught in the mid-1970s, did not meet the present needs of students nor those of the future. These landmark studies and others initiated by many individual and professional groups in the early 1980s have served as motivation for broad assessments of science education programs in schools across the nation during the decade that has followed.

Numerous recent articles have been critical of the standardized science test scores of American students when compared with the scores of their Asian and European counterparts. A concern for the direction of science education, the level of attainment by American science students, and the extent of science concept development have been noted by education professionals as well as members of the general public.

One of the more significant reform movements to evolve from these past efforts is the STS focus for teaching science that is described in the chapters preceding this section. The Iowa Chautauqua Program is an inservice program that enrolls classroom science teachers and assists them in using the STS approach in their teaching. The Iowa Chautauqua Program stimulates teachers to collect data as evidence for changes in outcomes of students who have experienced the STS approach to science. The publication *Chautauqua Notes* details results of assessments of data collected by STS teachers. One of the fears about STS is that science concept attainment by students who experience the STS approach will be decreased in favor of situations where concept mastery is emphasized as the primary indicator of learning. Of course, comparing concept attainment when students learn science via STS compared with traditional approaches is of great interest.

Recent research (Blunck and Yager, 1990; Mackinnu, 1991; McComas, 1988, 1989; Myers, 1988; Yager, Myers, Blunck, and McComas, 1990) indicates that students learn science best when given opportunities to construct for themselves the science concepts, processes, applications, attitudes, and creativity skills that permit them to understand their own environment. Students learn to satisfy their own perceived needs regarding how to come to equilibrium with numerous scientific/societal problems.

Successes with STS Instruction and Concept Mastery

Two major reports illustrate the effectiveness of the STS focus for teaching science (Mackinnu, 1990; Myers, 1988). These two research projects compared outcomes of students who experienced science taught with an STS focus with students who were taught with a more traditional textbook focus. Several teachers and several classes of students ranging from grade four to grade nine were compared and student outcomes were reported.

The results of student assessments within the concept domain produced significant results indicating that students do score well on examinations designed to measure concept mastery. Many science teachers object to the utilization of newer teaching techniques; they argue that students do not learn the science information as well as they do with more traditional methods. These new studies now clearly indicate that students do as well or better in STS-focused classrooms as do their traditionally taught peers.

Assessment data reported by STS teachers enrolled in the Iowa Chautauqua Program provide the following findings:

1. In 1988 it was reported that the pretest-posttest assessments of one class of eighth-grade STS students indicated a gain of sixteen points in mean scores when their pretest scores were compared to posttest scores (McComas, 1988).

2. In 1989 McComas reported comparisons of fifty-one classes of STS students (n = 1,100). Once again when pretest scores were compared to posttest scores gain scores of 33 percent were noted.
3. In 1990 Yager reported on the study of two science teachers involving fifty-two students. Pretest and posttest scores indicated that the STS students learned as many science concepts as did their traditionally taught counterparts.
4. In yet another study reported by McComas in 1989, 131 science teachers who had participated in Iowa Chautauqua Program used the pretest-posttest format to assess the effectiveness of the STS approach. Grades four through nine were involved in this assessment. Over 90 percent of the classes involved displayed positive changes in posttest scores with respect to gains in mastery of science concepts.

A Closer Look at Concept Mastery

Science teachers usually report concept mastery as a major goal of student attainment of their instruction. Such a goal of science teaching is shared by members of the general public. When asked to identify their perception of the most important student outcomes of school science, several groups of adults listed student learning of science concepts and science knowledge as among the most important objectives (Pogge and Yager, 1987). Judging from the structure and content of many American science textbooks, commercially prepared science curriculum guides, and standardized tests, many science teachers ascribe to such perceptions as does the American public.

Yager et al. (1990) have identified the following significant differences between STS and Traditional science teaching in terms of science concepts:

STS	*Traditional*
1. Students see concepts as personally useful	1. Concepts are really explanations of natural phenomena to be mastered for tests
2. Concepts are seen as a needed commodity for dealing with problems	2. Concepts are seen as outcomes in and of themselves
3. Concept learning occurs because of activity; it is an important happening but not a focus in and of itself	3. Learning of science concepts is principally for testing
4. Students who learn by experience retain concepts and can relate them to new situations	4. Retention of science concepts is short-lived.

Some Results in Muscatine, Iowa

Myers (1992) reported the following generalizations based on data available from comparisons of STS focused science programs to textbook-centered programs in ninth-grade science classes in Muscatine, Iowa:

1. Based on actual research data, it can be stated that science taught with an STS focus is effective in stimulating student learning of science concepts;
2. The STS approach to teaching science works equally well for all grade levels in stimulating learning of science concepts;
3. The STS approach works as effectively for students for long periods of time as it does for shorter periods of time;
4. STS students score equally well on standardized tests as do their traditionally taught counterparts;
5. STS works as well for males as for females;
6. STS approaches work for higher-achieving students as well as for lower achieving students; and
7. Students of teachers utilizing STS strategies for the first time in their science classrooms learn as many science concepts as do students in traditionally taught classrooms.

In a rigorously controlled experiment involving fifteen experienced STS teachers, Mackinnu (1991) studied differences in concept mastery for students in textbook-bound sections versus those in STS sections where the content focus (unit organizer) was the same. The same pre- and posttests were given to students in both treatments for each of the fifteen teachers.

Mackinnu found that there were no significant differences between any of the sections taught by the teachers on the pretest measures. Since a variety of content areas and grade levels were involved, the experiment is viewed as separate replications of the same experiment.

At the end of each of the fifteen experiments, posttests to measure concept mastery revealed no significant differences between any of the textbook and STS sections. The findings indicate that students master as many concepts of science in classes where the concepts are the organizer for the instruction as they do in STS sections where the concepts are studied as they emerge as needed information to resolve the problems, questions, and issues used as organizers for instruction. Students learn as many science concepts when they are encountered in real-world settings as when they are presented directly as content important to master for success in a course. The STS approach—though not focusing on concept mastery—results in as much conceptual learning as that which occurs in more traditional courses where concepts are identified directly for students to master.

Based on all data currently available, it is apparent that when the STS approach is utilized in the science classroom, the students learn science concepts as well, and in many instances better, than do students who experience science taught with the textbook-centered approach. However, setting mastery science concepts aside for a moment, there are even more compelling arguments favoring the STS approach to school science that are elaborated in chapters that follow.

References

Blunck, S. M., and Yager, R. E. (1990). The Iowa Chautauqua Program: A model for improving science in the elementary school. *Journal of Elementary Science Education*, *2*(2), 3–9.

Harms, N. C., and Yager, R. E. (eds.). (1981). *What research says to the science teacher*, Vol. 3. Washington, D.C.: National Science Teachers Association.

Helgeson, S. L., Blosser, P. E., and Howe, R. W. (1977). *The status of pre-college science, mathematics, and social science education: 1955–75*. Columbus, Ohio: Center for Science and Mathematics Education, Ohio State University.

Mackinnu. (1991). *Comparison of learning outcomes between classes taught with a science-technology-society (STS) approach and a textbook oriented approach*. Unpublished doctoral dissertation, University of Iowa, Iowa City.

McComas, W. F. (1988). Putting STS to the test. *Chautauqua Notes*, *3*(8), 2–3.

McComas, W. F. (1989). Just the facts: The results of the 1987–88 Chautauqua workshops. *Chautauqua Notes*, *4*(4), 1–2.

Myers, L. H. (1988). *Analysis of student outcomes in ninth grade physical science taught with a science/technology/society focus versus one taught with a textbook orientation*. Unpublished doctoral dissertation, University of Iowa, Iowa City.

Myers, L. H. (1992). STS and science concepts. In R. E. Yager (ed.), *The status of science-technology-society reform efforts around the world: ICASE yearbook 1992* (pp. 76–80). Hong Kong: International Council of Associations for Science Education.

Pogge, A. F., and Yager, R. E. (1987). Citizen groups' perceived importance of the major goals for school science. *Science Education*, *71*(2), 221–227.

Stake, R. E., and Easley, J. (1978). *Case studies in science education*, Vols. I and II (Stock No. 038-000-00364). Washington, D.C.: U.S. Government Printing Office, Center for Instructional Research and Curriculum Evaluation, University of Illinois at Urbana-Champaign.

Weiss, I. R. (1978). *Report of the 1977 national survey of science, mathematics, and social studies education: Center for educational research and evaluation.* Washington, D.C.: U.S. Government Printing Office.

Yager, R. E. (1990). Instructional outcomes change with STS. *Iowa Science Teachers Journal*, *27*(1), 2–20.

Yager, R. E., Myers, L. H., Blunck, S. M., and McComas, W. F. (1990). The Iowa Chautauqua program: What assessment results indicate about STS instruction. In D. W. Cheek (ed.), *Technology Literacy V: Proceedings of the Fifth National Technological Literacy Conference, Arlington, Virginia, February 2–4, 1990* (pp. 133–147). Columbus, Ohio: ERIC Clearinghouse for Science, Mathematics, and Environmental Education.

Chapter 6

Process Skills Enhancement in the STS Classroom

Julie Wilson
Sylvia Livingston

Process Skills and the Reforms

Science education has always articulated a need to have students develop their thinking and reasoning skills. The processes that scientists use have been seen as important to impart as "learnings" in science courses. Documentation of this focus for science teaching began as early as 1932, in the National Study of Education Yearbook (National Society for the Study of Education), to just recently with the *Benchmarks for Science Literacy: Project 2061* (AAAS, 1993) and the National Science Education *Standards* (NCSESA, 1993). Over time there have been a multitude of curriculum projects that have emphasized the process skills that scientists use for students to learn.

The 1960s were most noted for a form of teaching that emphasized the use of the special skills that scientists use as they do science. The programs for elementary science represented this approach the best; examples include: Science—A Process Approach (S-APA), Elementary Science Study (ESS), and Science Curriculum Improvement Project (SCIS). Students involved in these projects often demonstrated an improved performance over students found in traditional classrooms (Shymansky, Kyle, and Alport, 1983). Although programs at the middle and high school level were similar in design, the instructional framework lacked the openness found in the elementary programs. Specifically, Physical Science Study Curriculum (PSSC), Integrated Science Curriculum Study (ISCS), and Biological Science Curriculum Study (BSCS) frequently specified the problem and directed the course of the investigation (Herron, 1971). Aside from the lack of continuity in the curricular design, the programs of the 1960s resulted in improved student performance of science process skills, when compared to traditional textbook instruction (Kyle, 1982).

Projects of the 1990s revisit and utilize the processes of sciences identified by scientists and incorporated into reform programs of the 1960s. Most notable of these are the *Benchmarks for Science Literacy: Project 2061* (AAAS, 1993) and the National Science Education *Standards* (NCSESA, 1993). The *Benchmarks* describe these as "Habits of Mind" and "The Nature of Science." Although these reforms emphasize the processes of science, they neglect the same element that previous reforms have—the context for instruction. Specifically a context for learning that is relevant, seeks connections between concepts, shows science as a way of thinking, and works from the current conceptual understanding of students. One instructional movement is filling the context void, while stressing the importance of the processes in science.

The STS instructional movement provides a context for the use of processes skills used in science. The differences of the context between traditional process instruction and the STS process instruction has been contrasted by Yager and Roy (1993). Evident is the student as learner of science, instead of receiver of science, in a context in which there is an emphasis on process skills that students use as they resolve their own problems (NSTA, 1990–1991). Table 6.1 shows Yager and Roy's (1993) comparison.

TABLE 6.1 Process of Traditional versus STS

Traditional	*STS*
1. Science processes are skills scientists possess	1. Science processes are skills students themselves can use
2. Students see science processes as something to practice as a course requirement	2. Students see science processes as skills they need to refine and develop themselves more fully
3. Teacher emphasis on process skills not understood by students because these skills rarely contribute to actions outside of, or even to, the course	3. Students readily see the relationship of science processes to their own actions
4. Students see science processes as abstract, glorified, unattainable skills	4. Students see processes as a vital part of what they do in science classes

In previous reform initiatives, process science has been emphasized, but lacked the acknowledgment of the context for instruction. Students participated in the processes found in science, yet they had no bridge to other science or real-life experiences. The new reforms, especially STS, address the context in a meaningful way that is setting the agenda for other instructional programs. Integral to STS's contextual acknowledgment is constructivist theory (e.g., Brooks and Brooks, 1993; von Glasersfeld, 1988), which is discussed in Part II of this monograph. As a result students are learners using the processes found in science in a situation that is meaningful, relevant, and developmentally appropriate.

CONNECTING PROCESS SKILLS

Hurd (1986, 1991) has admonished that most high school students lack the intellectual skills in science and technology that allow them to assume their roles in society and that students need to cultivate scientific patterns of thinking, logical reasoning, curiosity, openness to new ideas, and skepticism in evaluation of claims and arguments. One is left to conclude that students need to utilize the process skills characterizing science in order that they can understand and produce their own knowledge, which will ultimately allow them to participate actively in society. This is especially critical when contemporary life is bombarded daily with issues and problems based in science and technology. The ability to use the processes learned in science classes should provide students with a mechanism to evaluate, contemplate, and react to the changing world today. Unfortunately, there is little evidence to indicate that science so serves most students.

As science educators attempt to conduct science as Hurd (1986, 1991) suggests, they are faced with decisions and considerations in using the processes characterizing science. Primarily, decisions and considerations that are tied to the "connectedness" of science process skills; now the processes connect to each other, the learner, and future problems, issues, or investigators. Thus, the "connectedness" should be addressed in a way that is developmentally appropriate, meaningful, and relevant to the learners. In doing this, science educators become guides or facilitators of instruction. This entails monitoring students as they proceed through a problem, issue, or investigation; providing information at critical moments; and asking questions that cause reflection on specific processes. Although appearing simplistic at first, the problem has a great deal of inherent complexity. Science educators could not possibly address all three levels of process connections in their instruction, unless instruction is centered around the student.

By using the student as the template for classroom instruction, science processes are integrated through their actions. The teacher does not have to explain to each student which processes they are experiencing. As students proceed through an investigation they select, they use the processes they need. The nature of the investigation blurs divisions of basic and integrated process skills. Basic process skills such as observing, classifying, communicating, measuring, using space/time relations, using numbers, and inferring and predicting are used in conjunction with the integrated skills of controlling variables, interpreting data, formulating hypotheses, defining operationally, and experimenting (skills defined by AAAS, 1965). Furthermore, using process skills in a combined state supports the development of thinking skills; an initial tenet of using process skills. Baird and Borich (1987) and Padilla, Okey, and Dillashaw (1983) have suggested that these skills are often viewed independently; yet they have been observed as correlating to one another and with Piaget's construct of formal reasoning.

Bridging both basic and integrated process skills into a student's repertoire of critical thinking has been a long-time concern of science educators. The final test of successful incorporation of a student would be the transfer of the skills to new and novel situations. Studies suggest that the longer the instructional periods of using the processes found in the science, the greater the gains in the students' use of these science process skills (Finley, Lawrenz, and Heller, 1992; Padilla, Okey, and Garrard, 1984). Specifically, Padilla, Okey, and Garrard (1984) found that students who used integrated process skills for an extended duration of time showed growth in identifying variables and stating hypothesis. Roth and Roychoudhury (1993) found that authentic contexts supported the use of the process skills used in science. Thus, both duration and context play a role in the students acquiring of basic and integrated science processes.

Once these processes are "incorporated" students will hopefully resort to them when they are faced with new problems or investigations. Perkins and Salomon (1989) referred to transfer in problem solving as being a high or low road to transfer. The "low road" depends on previous experience and similarity to previously practiced problems. The "high road" depends on the learner's ability to abstract the problem. The learner must reflect and retrieve previous information and apply it to the new situation. Ultimately the degree of transfer of process skills to new or novel situations may be dependent on previous experiences.

STS offers a "connectedness" to the process skills scientists engage in. Student select problems, issues, or investigations that are meaningful and relevant. As students engage in these investigations, they experience both basic and integrated process skills and assimilate and refine their repertoire of the skills found in science. With prolonged involvement of multiple STS opportunities students create avenues of transfer to new situations. These three areas provide the "connectedness" that is found in the science that Hurd (1986, 1991) promotes. Science is not an accumulation of knowledge about a particular domain, but competence in the use of process skills that are basic to all science (Roberts, 1982). In this way, students involved in an STS problem, issue, or investigation see the processes as a way of knowing and doing science. In the broader scheme, students are provided with the skills to evaluate, contemplate, and react to their world.

STS and Process Skills

STS advocates find that the goals and instruction characterizing STS are congruent with the needs of the student. Students pursue investigations at their own knowledge, cognitive, and interest level. Students work with current questions, problems, and issues that are important, relevant, and stated by them. As students act on these investigations in a manner that they elect to follow, they utilize the processes that are found in science. This context allows students to become internally motivated to use the processes found in science, reflective about the need of science processes, and refine the processes that have and will be used.

Successful practices for students require a positive affective domain within the instructional context. STS has been found to have a positive effect on students' attitudes toward science (McComas, 1993). The implication of this is far reaching when it comes to science instruction. STS provides a way to combat the negative feeling that students have had about science over the years (Jones, Mullins, Raizen, Weiss, and Weston, 1992; Mullins and Jenkins, 1988). An increase in the affective domain suggests that students find personal satisfaction in their participation in science. Students who enjoy and like science are more likely to engage in science (Simpson & Oliver, 1990). As students have an active engagement in STS, students create opportunities to use the process skills characterizing science. Positive feelings arising from STS instruction encourages a depth of process. Students then integrate processes into their investigations, refine them with repeated usage, and transfer them to other areas that require the same processes.

Research results have confirmed that student involvement with STS strategies are superior to typical textbook/laboratory instruction in stimulating growth in science process skills (Iskandar, 1991; Liu, 1992; Mackinnu, 1991; McComas, 1989). In past studies students across all grades who have been involved in STS gained a further understanding of all basic science process skills. An evaluation of process skills can occur at the level of the classroom or individual student skill within the class group. For example, Liu (1992) looked at fifteen teachers using both the STS approach and a traditional approach. Teachers who implemented the STS approach stimulated a significant difference ($P < 0.05$) in student mastery of science process skills. Every STS classroom was significantly different in obtainment of science process skills. The mean differences and T-values can be found in Table 6.2. Iskandar looked at the percentages of specific science processes that were demonstrated by students of seventeen teachers who conducted both STS and non-STS classrooms. She found that the differences in the mastery of science process skills were significantly different ($P < 0.01$) in favor of the STS approach. STS classrooms typically had process use that was twice what traditional classrooms had. Percentages in each skill area can be found in Table 6.3. Both studies used the Process Instrument from the *Iowa Assessment Handbook* (Tamir, Yager, Kellerman, and Blunck, 1991).

TABLE 6.2 A Comparison of Change in Student Mastery of Science Process Skills for Fifteen Teachers Using STS and Non-STS Teaching Strategies (Liu, 1992)

	STS			*Non-STS*		
Teacher	*n*	*Mean Difference*	*T-Values*	*n*	*Mean Difference*	*T-Values*
1	25	4.92	15.84	26	0.15	0.84
2	27	3.22	13.38	26	0.96	8.18
3	18	3.44	18.64	20	0.60	3.94
4	22	3.63	15.60	23	1.17	7.24
5	26	4.96	17.38	25	0.88	6.60
6	21	6.04	14.51	22	0.81	5.23
7	26	5.34	13.24	24	1.33	4.87
8	22	6.18	21.20	24	1.25	5.31
9	23	5.30	16.10	24	0.91	3.25
10	27	6.63	27.00	26	1.23	6.23
11	17	5.52	8.05	16	0.87	4.34
12	29	7.31	21.23	28	0.92	41.00
13	25	6.64	21.04	27	1.37	7.36
14	25	4.04	12.67	26	1.26	7.04
15	27	5.33	26.70	25	1.16	6.45

TABLE 6.3 A Comparison of Student Demonstration of Specific Science Process Skills in STS and Non-STS Classrooms (Iskandar, 1991)

	Percentage of Students Demonstrating Specific Skill	
Process Skill	*N = 199 STS*	*N = 160 non-STS*
Using space/time relations	52	12
Observing	83	30
Classifying	87	27
Interpreting data	89	30
Inferring	73	40
Communicating	89	39
Controlling variables	63	21
Drawing conclusions	85	23
Predicting	73	20
Using numbers	89	39
Measuring	90	32
Comparing and differentiating	87	32
Hypothesizing	64	18
Selecting best example and procedure	53	24

These studies support the idea that STS classrooms provide opportunities for students to refine and evolve the processes found in science. Students had a greater mastery and utilization of process skills in these classrooms as opposed to classrooms where traditional instructional methods were used. Furthermore, there was no demonstrated differences in the acquisition of processes between males and females.

Most science programs provide predetermined experiences that students need to use to learn process skills; the STS approach is a student-centered way where process skills are introduced and used only when needed. STS investigations are facilitated by a teacher in a context that allows the student to act on questions, problems, or issues of personal significance. This type of investigation, authentic and activity based, has consistently been shown to increase students' use and the acquisition of process integrated process skills (Padilla, Okey, and Garrard, 1984; Roth and Roychoudhury, 1993; Shymansky, Hedges, and Woodworth, 1990). Furthermore, STS exceeds the authenticity and activity parameters by adding the component of personal relevance and local ties. Hence, the use of science process skills that are unique to science and its understanding occur in situations where students have selected the question, problem or issues, and designed a path for investigation. This could theoretically contribute to the marked increase in students' knowledge of the process skills found in science classes where the STS approach is used.

An Example from the Class

An example of focusing on science process skills was found in the ninth grade physical science program at City High (Iowa City, Iowa). The chemistry unit began with students looking at the chemicals found in products. Students were surprised at the amount of preservatives, flavor enhancers, and the claims advertised by the products. As the students investigated more and more products, their initial concern became amplified and the classroom took a student-centered direction. Their concern ultimately developed into the decision to initiate a full scale "Consumer Reports" investigation of various products.

Students tested and evaluated products in-depth, based on identified concerns from the class and the general public. Products were assessed on environmental impact, advertising claims, and actual performance. A few of the tests included: strength testing of paper towels, soap residue on hands after washing, resistance to chipping of nail polish, and sogginess of breakfast cereal. Over fifty different types of products were tested, ranging from chocolate chips to ball point pens.

During the testing period the smells of popcorn, perfume, nail polish remover, soap, and sour cream potato chips could be noted. Procedures included field testing, laboratory testing, bulletin board surveys, random surveys, interviews, and observations. Students designed tests, collected data, both quantitatively and qualitatively, and interpreted the findings. Several final evaluations even had weighted categories, reflecting the importance of some areas over others.

The testing period also became a jumping-off point for several chemistry and physics lessons. Students wanted to know about (to name a few) pH, reactions, chemicals, acids, bases, and friction. The context of instruction provided content connections, while using the processes found in science. Students, through their investigations, created a need to know specific science content. At the conclusion of the testing, students expressed a desire to communicate their findings to others. They elected to set up and participate in an evening open house for the parents, fellow classmates, school personnel, district personnel, and the public. Products and their testers were located in the cafeteria and the hallways of City High to answer questions, discuss findings, and provide information to over 150 visitors; including representatives from the local media.

The following day the students expressed how rewarding it was to pass on their information and how often consumers appreciated finding out what were the best products—even if it did go against their current product preference. Students also expressed the value of learning science in a meaningful and relevant situation which they chose.

From the onset, students saw a problem, took ownership, and acted on it. Students developed tests for products and evaluated their findings. But beneath the action the students were challenged to redefine their knowledge base and their skills with science processes used in doing the investigation. Students were allowed and encouraged to experience thinking found in science in a situation they directed.

The Power of Process Skills

Using the process skills found in science in an STS context, students have experiences that are meaningful, relevant, and developmentally appropriate. In addition, they find a way of knowing about questions, problems, and issues that surround them. This type of investigation exceeds the traditional process classroom instruction. To do this the STS framework encourages:

1. a meaningful context that has students building their own knowledge base;
2. complex challenges that have students using basic and integrated processes of science;
3. reflection about the consequences of procedural decisions;
4. continuing development of higher-order thinking skills; and
5. opportunities for transfer of processes to new and novel situations.

Students involved in STS classrooms have shown significant gains in the use of process skills. Essential to these gains is the positive affective responses of students and the context in which STS is implemented. Students enjoy the science in which they are involved (McComas, 1993). The established context is central to this. Students direct their learning and place it in a situation that is meaningful, relevant, and developmentally appropriate to them.

As students experience the process skills found and practiced in science, they see these processes as integrating one to another and not as isolated skills teachers or curriculum materials purport to teach. Basic and complex skills intertwine with each other throughout student investigations. Students also find that the skills become assimilated into a framework that facilitates a way of knowing. As they experience the process skills in science in authentic contexts, the skills become part of a repertoire of knowledge. With repeated experiences, students can demonstrate the transfer of skills to situations that are novel and new. Process skills offer a "connectedness" that promote a way of knowing.

STS and its utilization of the process skills in science succeed where reforms in the past have failed. STS instruction takes the processes of science and provides a mechanism for the meaningful integration with science concepts. As students participate they develop a "way of knowing" and a "desire to know." This is truly unique in the world of reforms.

REFERENCES

American Association for the Advancement of Science. (1965). *The psychological basis of science—A process approach.* Washington, D.C.: Commission on Science Education, Author.

American Association for the Advancement of Science. (1993). *Benchmarks for science literacy: Part I: Achieving science literacy: Project 2061.* Washington, D.C.: Author.

Baird, W. E., and Borich, G. D. (1987). Validity consideration for research on integrated science process skills and formal reasoning ability. *Science Education, 71*(2), 259–269.

Brooks, J. G., and Brooks, M. G. (1993). *In search of understanding: The case for constructivist classrooms.* Alexandria, Va.: Association for Supervision and Curriculum Development.

Finley, F., Lawrenz, F., and Heller, P. (1992). A summary of research in science education-1990. *Science Education, 76*(3), 239–254.

Herron, M. (1971). The nature of scientific inquiry. *School Review, 79,* 171–212.

Hurd, P. DeH. (1986). Perspectives for the reform of science. *Phi Delta Kappan, 67*(5), 353–358.

Hurd, P. DeH. (1991). Why we must transform science education. *Educational Leadership, 49*(2), 33–35.

Iskandar, S. M. (1991). *An evaluation of the science-technology-society approach to science teaching.* Unpublished doctoral dissertation, University of Iowa, Iowa City.

Jones, L. R., Mullis, I. V. S., Raizen, S. A., Weiss, I. R., and Weston, E. A. (eds). (1992). *The 1990 science report card: NAEP's assessment of fourth, eighth, and twelfth graders.* Washington, D.C.: Education Information Branch of the Office of Education Research and Improvement.

Kyle, W. C., Jr. (1982). *A meta-analysis of the effects on student performance of new curricular programs developed in science education since 1955.* Unpublished doctoral dissertation, University of Iowa, Iowa City.

Liu, C. (1992). *Evaluating the effectiveness of an inservice teacher education program: The Iowa Chautauqua Program.* Unpublished doctoral dissertation, University of Iowa, Iowa City.

Mackinnu. (1991). *Comparison of learning outcomes between classes taught with a science-technology-society (STS) approach and a textbook oriented approach.* Unpublished doctoral dissertation, University of Iowa, Iowa City.

McComas, W. F. (1989). Science process skills in STS education: The results of the 1987–88 Chautauqua workshops. *Chautauqua Notes, 4*(7), 1–3.

McComas, W. F. (1993). STS education and the affective domain. In R. E. Yager (ed.) *What research says to the science teacher, Volume 7: The science, technology, society movement* (pp. 161–168). Washington, D.C.: National Science Teachers Association.

Mullis, I. V. S., and Jenkins, L. B. (eds.). (1988). *The science report card: Elements of risk and recovery.* Report Number 17-S-01. Princeton: Educational Testing Service.

National Committee on Science Education Standards and Assessment. (1993). *National science education standards: July '93 progress report.* Washington, D.C.: National Research Council.

National Science Teachers Association. (1990–1991). Science/technology/society: A new effort for providing appropriate science for all (Position Statement). In *NSTA Handbook* (pp. 47–48). Washington, D.C.: Author.

National Society for the Study of Education, Committee on the Teaching of Science. (1932). *Thirty-first Yearbook: A program for teaching science: Part I.* Chicago: The University of Chicago Press.

Padilla, M. G., Okey, J. R., and Dillashaw, F. (1983). The relationship between science process skills and formal thinking abilities. *Journal of Research in Science Teaching, 20*, 239–246.

Padilla, M. G., Okey, J. R., and Garrard, K. (1984). The effects of instruction on integrated science process skill achievement. *Journal of Research in Science Teaching, 21*(3), 277–287.

Perkins, D. N., and Salomon, G. (1980). Are cognitive skills context bound. *Educational Researcher*, *18*, 16–25.

Roberts, D. (1982). Developing the concept of "curriculum emphasis" in science education. *Science Education, 66*(2), 243–260.

Roth, W. M., and Roychoudhury, A. (1993). The development of science process skills in authentic context. *Journal of Research in Science Teaching, 30*(2), 127–152.

Shymansky, J. A., Hedges, L. V., and Woodworth, G. (1990). A reassessment of the effects of inquiry-based science curricula of the 60s on student performance. *Journal of Research in Science Teaching, 27*(2), 127–144.

Shymansky, J. A., Kyle, W. C., and Alport, J. M. (1983). The effects of new science curricula on student performance. *Journal of Research in Science Teaching, 20*(5), 387–404.

Simpson, R. D., and Oliver, J. S. (1990). A summary of the major influences on attitude toward and achievement in science among adolescent students. *Science Education, 74*(1), 1–10.

Tamir, P., Yager, R. E., Kellerman, L., and Blunck, S. M. (1991). *The Iowa assessment handbook.* Iowa City: University of Iowa, Science Education Center.

von Glasersfeld, E. (1988). *Cognition, construction of knowledge, and teaching.* Washington, D.C.: National Science Foundation.

Yager, R. E., and Roy, R. (1993). STS: Most persuasive and most radical of reform approaches to "science education." In R. E. Yager (ed.) *What research says to the science teacher, Volume 7: The science, technology, society movement* (pp. 7–13). Washington, D.C.: National Science Teachers Association.

CHAPTER 7

THE AFFECTIVE DOMAIN AND STS INSTRUCTION

William F. McComas

ENLARGING THE DOMAINS FOR ASSESSING SUCCESS?

Those concerned with the impact of school on students have recently come to realize that the form and focus of schemes used for assessing the progress of individual students, schools, and curriculum reform efforts are generally lacking. In spite of calls to widen the assessment agenda from various corners of the science education community (Klopfer, 1976*a*; Yager and McCormack, 1989), school tests continue to be narrowly focused on assessment of intellectual skills such as concept acquisition, retention and, only occasionally, the ability to apply what has been learned. Other areas such as the affective domain (Krathwohl, Bloom, and Masia, 1956)—that of feelings and values—have never received much emphasis in classroom assessment. This chapter will highlight the affective domain, its value in science education, and the impact on the attitude domain made by STS instructional orientations.

WHAT CONSTITUTES THE AFFECTIVE DOMAIN IN SCIENCE EDUCATION?

A major aspect associated with the affective domain is that of its definition and delineation. A wide number of researchers (Aiken and Aiken, 1969; Curtis, 1924; Gardner, 1975*a*; Gauld and Hukins, 1980; Haney, 1964; Klopfer, 1976*b*; Koballa and Warden, 1992; Mulkay, 1979; Munby, 1983) have proposed elements to be included within the region of science attitudes. From their work it is possible to extract several broad attitude-related areas including preferences, values, acceptance, beliefs, appreciation, interest, opinions, and commitment. In addition to these attributes of the attitudinal realm, some theorists include scientific attitude items such as curiosity, weighing evidence, skepticism, objectivity, and suspension of judgment within the affective domain (Germann, 1988; Jones and Butts, 1983). In these cases, care must be taken not to confuse attitudes *toward* science, science classes, and science teachers with *scientific attitudes*. Broad statements about "student attitudes" toward anything should be viewed with suspicion until one investigates just what attitudes are being measured and discussed. The review of the literature with respect to attitudes and STS presented here is restricted to those attitudes that would best be classified as opinion rather than scientific attitudes.

Why Should Science Educators Be Concerned with Student Attitudes?

When Schwab (1962) suggested that student support for inquiry outside of school is related to the development of favorable attitudes toward science in school, he stated what must be the strongest intuitive justification for concern with student attitudes. This personal view of Schwab has been substantiated empirically by the work of various researchers.

An interesting finding by Alderson and Walberg (1967) and Walberg, Singh, and Rasher (1977) links cognitive growth in twelfth grade physics to student perceptions that the class is well organized by the teacher and that students are provided freedom to question and learn in an informal atmosphere. Bloom (1976) reported correlations between student interest and science achievement at $r = +0.35$ for eighth graders and $r = +0.52$ for twelfth-grade students. This finding is substantiated by Hough and Piper (1982) who demonstrated a strong relationship ($r = +0.45$) between gain scores on science achievement tests and scores on an attitude inventory. In addition, Bloom (1976) found that approximately 23 percent of the variance in science achievement is linked to students' self-concept and opinions of what they were studying. A more recent study by Oliver and Simpson (1988) shows that affective variables accounted for approximately 20 percent of eleventh-grade chemistry achievement and 30 percent for twelfth graders.

In addition to the developing causative link between attitude and achievement, evidence is accumulating that student attitudes can be changed—perhaps inadvertently—by instructors and instruction. The area of attitude change in science is quickly gaining a theoretical base with the work of investigators such as Shrigley and Koballa (1987) building on the earlier investigations of Hovland, Janis, and Kelley (1953).

It is doubtful that any of these findings are surprising to experienced educators who likely acknowledge a strong relationship between attitude and science achievement. What is surprising and alarming is the growing body of evidence supporting the conclusion that, in general, the longer students study science, the less they like it. Declines in student attitudes toward science have been documented by Disniger and Mayer (1974), Randall (1975), Lawrenz (1976), Haladyna and Shaugnessey (1982), and by the past several National Assessments of Educational Progress (NAEP). In the summary of the 1986 NAEP study, Mullis and Jenkins (1988) state that "most students . . . appear to be unenthusiastic about the value and personal relevance of their science learning and their attitudes seem to decline as they progress through school" (p. 132). The trend is seen to continue with the results of the 1990 NAEP investigation. Students liked science less as they progressed through school. Eighty percent of fourth graders said that they liked science, 68 percent of eighth-grade students responded affirmatively, but only 65 percent of twelfth graders indicated that they liked science (Jones, Mullis, Raizen, Weiss, and Weston, 1992). Simpson and Oliver (1990) substantiate the national finding with their examination of the same group of students through several years. They found that both attitude and motivation with respect to science achievement fell off from the start of the school year to the end and also declined from one year to the next.

Unfortunately, from both the perspective of assessment and instruction, teachers do not seem to value affective objectives as highly as they do the more traditional cognitive ones (Gardner, 1975*b*; Schibeci, 1977, 1981). Teachers typically do not assess how students feel about what they are learning and how they are being taught. This may be due to a lack of knowledge about the affective domain or perhaps because teachers do not know what use to make of the results gleaned from an evaluation of student attitudes. Bloom, Hastings, and Madaus (1971), developers of the taxonomy of the affective domain, suggest that teachers may even deliberately choose not to explore the affective domain and avoid altering or clarifying student attitudes because of fear that they will be accused of indoctrinating their students.

Given the realization that students do indeed form attitudes about the discipline and instruction in spite of other goals held by their teachers, it is reasonable to characterize much of the affective domain as part of the hidden or unstated curriculum. As such, educators should be concerned about the attitudes students are forming as a result of experience in school particularly since positive student attitudes toward science translate into higher levels of science achievement.

STS Education and the Affective Domain

Worldwide, STS as an instructional organizer has taken on various forms with differing objectives, but most STS reform efforts are characterized by the presence of two intertwined components. STS programs typically highlight the impact of science on society and involve students in their own learning at a much higher degree than has been the case in other reform plans. The high degree of student involvement make it quite likely that STS instructional schemes will impact student attitudes—both about science itself and science instruction—and likely do so in a positive fashion.

Not only is STS instruction likely to impact the affective domain, but even the stated goals for STS instruction typically include goals targeting student attitudes. The NSTA (1982, 1990), for instance, characterizes STS programs as those that have, among other attributes, student identification of problems, students actively involved in seeking information applied to solve real-life problems, extension of learning beyond the classroom, impact of science and technology on individual students, opportunities for students to experience citizenship roles and autonomy in the learning process. Yager and Roy (1993), major proponents of STS education, contrast the goals of traditional science education with STS education on a number of counts including the view that students should actively seek information, enjoy the science experience, increase their interest in science from year to year, become more curious about science, and see their teachers as facilitators and guides rather than ultimate information sources. The STS view stands in direct opposition to the view that mathematics and science classrooms too often promote a passive role for students, depending heavily on textbooks, teacher talk, and worksheets. In general, the stated curriculum does not foster complex thinking or work on authentic problems (O'Neill, 1991).

What Research Says about STS and Student Attitudes

STS as a teaching orientation continues to gain support from science educators, science teachers, and textbook publishers, but its effects on any of the domains of student learning have rarely been measured in any organized and controlled fashion. As indicated in the introductory sections of this chapter, the realm of student attitudes toward science, science classes, and science teachers has been of interest to educational theorists for some time, but only in the past few administrations of the National Assessment of Education Progress (NAEP) has this facet of the school experience become more widely measured.

The attitude items included on the NAEP have been modified slightly and included in a number of appraisals of the effect of an STS teaching orientation on student attitudes in a number of studies. The Iowa Science Attitude Inventory (McComas and Yager, 1989) instrument developed for use in the appraisal of student attitudes has as its central objective the assessment of student perceptions of their science classes, science teachers, and the value of science study. The Inventory consists of approximately fifty items with the choice of "Yes," "No," and "I Don't Know" provided for questions that appear in a form such as, "Does your teacher enjoy science?" In both a small scale pilot study (McComas, 1989) and a large, statewide, matched-pairs study (McComas, 1993), student attitudes were seen to change significantly and positively when students engaged in an issues-oriented mode of study. In all cases, the students were in classes taught by teachers who attended a series of inservice workshops designed to communicate the basic tenets and methods of STS instruction. Table 7.1 provides a summary of the results of the statewide STS study (McComas, 1993).

TABLE 7.1 Categories of Question Types and Significant Changes from Pretest to Posttest on the Iowa Science Attitude Inventory in Classes Taught with an STS Orientation Matched Pairs—Grades 1 and 3–12 (N = 1735)

General Category of Questions		*Number and Percent of Questions in Category Showing Change from Pretest to Posttest*	
Questions about attitudes with respect to . . .	*Number of Questions in Category*	*Number of Questions Showing Change*	*Percent Showing Change*
School in general	3	0	0%
Usefulness of science knowledge	5	4	88%
Technology	1	1	100%
Science teachers	12	9	75%
Being a scientist	6	2	33%
Solving world problems	8	4	50%
Science class	13	6	46%
Study of science	4	4	100%

The data from this survey indicate that students seem increasingly more positive when asked questions about the usefulness of science information outside of school, the applicability of science class to future careers, the place of technology in daily living, and basic feelings about their experiences with school science.

In the category of student perceptions of the value of studying science, students find science information increasingly more useful outside of school, more useful for further study, and more helpful with career choices than they did before STS instruction. At the same time, students feel that science class is exciting and fun. Students also agree that technology affects daily living; this change in perception clearly illustrates that the technology aspect of STS has been communicated well.

STS units of instruction caused increases in the way students perceive their teacher, and perhaps in the way in which teachers themselves perform. As found in an earlier study (McComas, 1989), the greatest change is in the area of whether students felt teachers admitted to not knowing all the answers to questions. It is clear that STS teachers have successfully communicated the idea that they *cannot*, and should not, be considered the sole source for all information in their students' lives. Approximately 60 percent of the students perceived that their teachers admit to not knowing answers to all student questions before the introduction of the STS unit, but after the issue-based module of instruction, 71 percent of students felt that teachers did not know all the answers. When teachers adopt the philosophy that issues, not just correct answers, are the real concern, this is no surprise.

Experimental Studies: A Meta-analysis

Although results such as those from surveys are encouraging and provide clues that an STS teaching orientation does make student attitudes more positive, the nature of descriptive studies does not permit firm conclusions linking instruction itself to improvements in attitude. What is necessary to make conclusive statements about the role of STS instruction in student attitude change are actual investigations in which students in STS and non-STS classes are compared.

To blunt criticisms that typical comparison studies are poorly designed, the literature was consulted to locate studies in which highly skilled STS teachers were asked to teach several sections of students using the STS approach and several classes with the more traditional textbook lecture-discussion approach in an experimental or quasi-experimental design. Three studies were located meeting the above criteria (Iskandar, 1991; Lu, 1993; Mackinnu, 1991), thus providing an opportunity to investigate suggestions that STS instruction can change student attitudes more significantly than does traditional instruction.

All three studies used an eighteen- or thirty-item attitude inventory with statements such as "science lessons are fun," "science lessons bore me," "a job as a scientist would be interesting." Each item includes a five-point Likert scale from strongly agree to strongly disagree. The questions are structured so that a choice of strongly agree, for instance, is always an indication of positive attitudes. Furthermore, student responses may be readily converted to a numerical result such that higher average values for a student represents generally more positive attitudes. This approach is useful for providing a single attitude score for each class and, of course, for performing statistical analysis, but by combining the data from a number of questions, the fine detail provided by a question-by-question analysis is lost.

Using the meta-analytic techniques suggested by Hedges, Shymansky, and Woodworth, 1989) and Hedges and Olkin (1985), effect sizes were calculated for each grade level, for each study across grade levels where possible, and for all three studies combined. This technique permits the formation of conclusions based on a much larger sample size than any single study permits. According to Hedges, Shymansky, and Woodworth (1989), the effect size is the "difference between the population mean criterion score for (the STS classes) and traditional curricula expressed in standard deviation units" (p. 23). Therefore, "an effect size of 1.0 indicates that a pupil who would have been at the mean, or fiftieth percentile, under the old curriculum would be one standard deviation above the mean . . . under the new (or STS) curriculum" (p. 23).

In the Iskandar (1991) study, twelve teachers, three at each grade level from sixth to ninth grades, well acquainted with the STS methodology were asked to teach one class with the traditional textbook method and one with the STS orientation. The results are summarized in Table 7.2. Both classes addressed the same content and lasted the same length of time. In this study several attitude measures were administered before and after the units of instruction. One measure, equivalent to those used in the other studies reported here, was an eighteen item Likert scale examining student attitudes about science class and scientists. A related attitude measure targeting student perceptions of their science teachers was reported on a question-by-question basis and hence was not appropriate for meta-analysis.

TABLE 7.2 A Summary of the Responses of Students to an Attitude Inventory Comparing Students Instructed with a Traditional Orientation and Those with an STS Orientation.

Grade and Group	*Number of Students*	*Mean Pretest*	*s*	*Mean Posttest*	*s*	*t*	*p*	*Effect Size*
6th Text	81	12.51	2.1	11.42	2.6	3.69	0.00*	3.34
6th STS	77	12.35	2.2	20.10	3.0	–28.73	0.00*	
7th Text	76	10.57	2.7	10.41	3.2	0.53	0.00*	2.39
7th STS	77	10.90	2.8	18.10	3.8	–27.02	0.00*	
8th Text	75	10.63	2.7	10.09	2.8	2.70	0.01	2.86
8th STS	76	11.16	2.8	18.11	3.8	–22.58	0.00*	
9th Text	76	11.71	2.2	11.08	2.7	2.95	0.00*	3.08
9th STS	71	11.65	2.3	19.42	3.3	–25.94	0.00*	

* Significant at $p < 0.05$

† *T*-tests were conducted between mean pretest and mean posttest for each group (Iskandar, 1991).

On the measure of student perceptions of their science teachers, Iskandar (1991) reports that there are very few significant differences between the STS and textbook groups at the pretest administration with no pattern discernible from grade to grade. However, at the posttest level, at every grade level, each of the twenty questions is answered in a significantly different fashion when the experimental and control groups are compared. What this means is that STS students have significantly different and more positive attitudes with respect to their science teachers on issues such as their level of interest in the students, the frequency with which the teacher admits to not knowing the answers to all questions, and the number and quality of questions asked of students.

Although instruction seems to have caused a significant change in student attitudes in both the textbook and STS groups (with the exception of the eighth graders). Careful examination of the data will show that attitudes decreased in those students in classes with the textbook orientation while becoming more positive in the STS classes. Furthermore, the effect sizes illustrate that the STS group had a significant advantage in the generation of positive attitudes. The pooled effect size calculated for all students in all grades was 2.87, which means that the average student in the STS group had attitudes almost three standard deviations more positive than if they had been in the textbook group.

Lu (1993) used a thirty-item attitude inventory that included questions addressing the students impressions of science teachers, science classes, and science generally. The questions in the Likert scale were designed so that as student attitudes become more positive, the numerical average on the instrument becomes larger. Lu involved both experienced and inexperienced STS instructors to teach two classes each, one with a traditional orientation and one with an STS focus. The results summarized in Table 7.3, include only the results from the classes of experienced teachers so that these findings may be compared with the other studies which also involved highly skilled teachers.

There were no significant differences between students at the pretest level, but as seen in Table 7.3, student attitudes become significantly more positive in all four grades (4–7) with STS instruction when compared with the control group.

Table 7.4 contains the pertinent data from Mackinnu (1991) who used the same thirty-item attitude instrument as applied in the Lu (1993) study. Mackinnu engaged fifteen teachers, 5 in grades 4 and 5, 5 in grades 6 and 7, and 5 in grades 7 and 8 to participate in the study. As in the other studies, these experienced STS instructors taught two classes, one with an STS focus and another with a traditional textbook orientation. As with the other studies reported, the pretest scores of the treatment and control groups showed no significant difference on the attitude measure, but when the posttests were administered, there were significantly more positive attitudes among those in the STS group, but no such increases were found in the traditional classes. "*T*-tests conducted on posttest scores at the end of [the] experiment indicated that all of them are

TABLE 7.3 A Summary of the Responses of Students to an Attitude Inventory Comparing Students Instructed with a Traditional Orientation and Those with an STS Orientation

Grade and Group	*Number of Students*	*Δ Mean Prepost*	*s*	*df*	*t*	*p*	*Effect Size*
4th Text	72	–1.79	0.72	116.3	–10.30	0.00*	1.74
4th STS	69	–0.15	1.11				
5th Text	47	–1.65	0.62	91.0	–9.32	0.00*	1.93
5th STS	46	–0.32	0.74				
6th Text	45	–1.35	0.67	89.0	–7.26	0.00*	1.52
6th STS	46	0.20	0.20				
7th Text	74	–1.38	0.76	142.0	–10.92	0.00*	1.82
7th STS	70	0.11	0.87				
4–7 Text	238	–1.55	0.73	436.4	–18.56	0.00*	1.71
4–7 STS	231	–0.11	0.93				

* Significant at $p < .05$

† *T*-tests were conducted between mean posttest for each group (Lu, 1993).

TABLE 7.4 A Summary of the Responses of Students to an Attitude Inventory Comparing Students Instructed with a Traditional Orientation and Those with an STS Orientation*

Grade and Group	*Number of Students*	*Mean Posttest*	*s*	*t*	*Effect Size*
4–9 Text	362	13.42	4.02	5.75†	1.62
4–9 STS	360	20.00	3.90		

* Significant at $p < 0.05$

† *T*-tests were conducted between mean posttest for each group. Data relative to this study have been extracted from Mackinnu (1991) and Yager and Tamir (1992).

significantly different either at $\alpha = .05$ or $\alpha = .01$. This means that the STS approach does offer an advantage over the traditional textbook oriented approach in the attitude domain. More important, the results are homogeneous or consistent across the fifteen pairs of classes or teachers. This homogeneity certainly gives additional confidence that the result typically holds across different independent implementations" (Mackinnu, 1991, pp. 62–63).

CONCLUSIONS

These studies, as summarized in Table 7.5, are conclusive: an STS orientation does make significant changes in student attitudes. Furthermore, those attitudes become resoundingly more positive. This analysis of student attitudes with respect to STS instruction has used a meta-analytical approach to summarize three quasi-experimental studies individually and collectively. These studies show effect sizes revealing that on average, a student in an STS class will likely have final student attitudes approximately two standard deviations more positive than those in traditional textbook classes.

TABLE 7.5 A Summary of the Three Studies Examined in this Meta-Analysis of the Effect of STS Instruction on Student Attitudes toward Science, Science Teachers, and Science Class

Researcher	*Number of Items on Instru-ment*	*Grades*	*Number of Tchrs STS*	*Number of Studts STS*	*Number of Tchrs Text*	*Number of Studts Text*	*TTL Studts*	*Overall Effect Size of Study*
Iskandar (1991)	18	6–9	12	301	12	308	609	2.87
Mackinnu (1991)	30	4–9	15	360	15	362	722	1.62
Lu (199)	30	4–7	10	231	10	238	469	1.71
Total	—	4–9	37	892	37	908	1,800	2.17

The earliest descriptive studies of the effect of STS on student attitudes were tantalizing because they seemed to show that students in STS classes had more positive attitudes after instruction than they did before such instruction. These studies were also frustrating because the nature of a pretest-posttest only investigation does not permit a firm link between the mode of instruction and apparent results.

The proponents of STS as an instructional organizer have suggested a number of goals for such instruction within the affective domain. While it remains to be seen if each attitude goal is similarly affected by STS instruction, a general conclusion regarding STS teaching and a variety of elements associated with the affective domain and science education is assured. The three studies cited in this analysis each illustrate clearly that there were no differences between the control and experimental groups before instruction and significant differences between the two groups of students after instruction with the STS student favored in the development of favorable attitudes toward science. The effect size calculations for both the individual studies and for all of the investigations taken together firmly address the assertion that STS instruction produces positive student attitudes toward science, science teachers, and science class.

References

Aiken, R. L., and Aiken, D. R. (1969). Recent research on attitudes concerning science. *Science Education, 53*, 295–305.

Anderson, G. J., and Walberg, H. J. (1967). *Classroom climate and group learning.* (ERIC Document Reproduction Service No. 015156.)

Bloom, B. S. (1976). *Human characteristics and school learning*. New York: McGraw-Hill.

Bloom, B. S., Hastings, J. T., and Madaus, G. F. (eds.). (1971). *Handbook of formative and summative evaluation of student learning*. New York: McGraw-Hill.

Curtis, F. D. (1924). Some values derived from researching general science. *Contributions to Education, 163*. New York: Columbia University, Teacher's College Press.

Disniger, J. F., and Mayer, V. J. (1974). Student development in junior high school science. *Journal of Research in Science Teaching, 11*(2), 149–155.

Gardner, P. L. (1975*a*). Attitudes to science: A review. *Studies in Science Education, 2*(1), 1–41.

Gardner, P. L. (1975*b*). Science curricula and attitudes in science: A review. *The Australian Science Teachers Journal, 21*(2), 23–40.

Gauld, C. F., and Hukins, A. A. (1980). Scientific attitudes: A review. *Studies in Science Education, 7*, 129–161.

Germann, P. J. (1988). Development of the attitude toward science in school assessment and its use to investigate the relationship between science achievement and attitude toward science in school. *Journal of Research in Science Teaching, 25*(8), 689–703.

Haladyna, T., and Shaugenssey, J. (1982). Attitudes toward science a quantitative synthesis. *Science Education, 66*(4), 547–563.

Haney, R. E. (1964). The development of scientific attitudes. *The Science Teacher, 31*(8), 33–35.

Hedges, L. V., Shymansky, J. A., and Woodworth, G. (1989). *Modern methods of meta-analysis*. Washington, D.C.: National Association of Science Teachers.

Hedges, L. V., and Olkin, I. (1985). *Statistical methods for meta-analysis*. New York: Academic Press.

Hough, L. W., and Piper, M. K. (1982). The relationship between attitudes toward science and science achievement. *Journal of Research in Science Teaching, 19*(1), 33–38.

Hovland, C. I., Janis, I. L., and Kelly, H. H. (1953). *Communication and persuasion.* New Haven: Yale University Press.

Iskandar, S. M. (1991). *An evaluation of the science-technology-society approach to science teaching.* Unpublished doctoral dissertation, University of Iowa, Iowa City.

Jones, B., and Butts, D. (1983). Development of a set of scales to measure selected scientific attitudes. *Research in Science Education, 13*(1), 133–140.

Jones, L. R., Mullis, I. V. S., Raizen, S. A., Weiss, I. R., and Weston, E. A. (eds.) (1992). *The 1990 science report card: NAEP's assessment of fourth, eighth and twelfth graders.* Washington, D.C.: Education Information Branch of the Office of Education Research and Improvement.

Klopfer, L. E. (1976*a*). Evaluation of learning in science. In B. S. Bloom, J. T. Hastings, and G. F. Madaus (eds.), *Handbook of formative and summative evaluation of student learning* (pp. 559–641). New York: McGraw-HIll.

Klopfer, L. E. (1976*b*). A structure of the affective domain in relation to science education. *Science Education, 60,* 299–312.

Koballa, T. R., and Warden, M. A. (1992). Changing and measuring attitudes in the science classroom. In F. Lawrenz, K. Cochran, J. Krajcik, and P. Simpson (eds.), *Research matters to the science teacher* (pp. 75–83). Washington, D.C.: National Association for Research in Science Teaching.

Krathwohl, D. R., Bloom, B. S., and Masia, B. B. (1956). *Taxonomy of educational objectives: Handbook II: The affective domain.* New York: David McKay.

Lawrenz, F. (1976). Student perception of classroom learning environments in biology, chemistry and physics courses. *Journal of Research in Science Teaching, 13*(4), 315–323.

Lu, Y. (1993). *A study of the effectiveness of the science-technology-society approach to science teaching in the elementary school.* Unpublished doctoral dissertation, University of Iowa, Iowa City.

Mackinnu. (1991). *Comparison of learning outcomes between classes taught with a science-technology-society (STS) approach and a textbook orientated approach.* Unpublished doctoral dissertation, University of Iowa, Iowa City.

McComas, W. F. (1989). Changing student attitudes with STS education. *Chautauqua Notes, 5*(1), 1–3.

McComas, W. F. (1993). STS education and the affective domain. In R. E. Yager (ed.), *What research says to the science teacher, Volume 7: The science, technology and society movement* (pp. 161–168). Washington, D.C.: National Science Teachers Association.

McComas, W. F., and Yager, R. E. (1988). *The Iowa assessment package for evaluation in five domains of science education.* Iowa City: University of Iowa, Science Education Center

Mulkay, M. (1979). *Science and the sociology of knowledge*. London: Allen and Unwin.

Mullis, I. V., and Jenkins, L. B. (eds.). (1988). *The science report card: Elements of risk and recovery*. Report Number 17-S-01. Princeton: Educational Testing Service.

Munby, H. (1983). Thirty studies involving the "scientific attitude inventory": What confidence can we have in this instrument? *Journal of Research in Science Teaching*, *20*(2), 141–162.

National Science Teachers Association (NSTA). (1982). *Science/Technology/Society (STS): Science education for the 1980s*. Position Paper. Washington, D.C.: Author.

National Science Teachers Association. (1990–1991). Science/technology/society: A new effort for providing appropriate science for all (Position Statement). In *NSTA Handbook* (pp. 47–48). Washington, D.C.: Author.

O'Neill, J. (1991). *Raising our sights: Improving U.S. achievement in math and science*. Alexandria, Va.: Association for Supervision and Curriculum Development.

Oliver, J. S., and Simpson, R. D. (1988). Influences of attitude toward science, achievement motivation, and science self concept on achievement in science: A longitudinal study. *Science Education*, *72*(2), 143–155.

Randall, R. R. (1975). A study of the perceptions and attitudes of secondary school students toward science as a school subject, science content and science teaching. *Dissertation Abstracts International*, *35*(3), 5152a.

Schibeci, R. A. (1977). Attitudes to science: A semantic differential instrument. *Research in Science Education*, *7*, 149–155.

Schibeci, R. A. (1981). Do teachers rate science attitude objectives as highly as cognitive objectives? *Journal of Research in Science Teaching*, *18*(1), 69–72.

Schwab, J. J. (1962). *The teaching of science as inquiry*. In J. J. Schwab and P. F. Brandwein (eds.), *The teaching of science* (pp. 3–103). Cambridge, Mass.: Harvard University Press.

Shrigley, R. L., and Koballa, T. R. Jr. (1987). *A theoretical framework: A decade of attitude research in science education*. University Park, Pa.: Pennsylvania State University, Department of Curriculum and Instruction.

Simpson, R. D., and Oliver, J. S. (1990). A summary of major influences on attitude toward and achievement in science among adolescent students. *Science Education*, *74*(1), 1–10.

Walburg, H. J., Singh, P., and Rasher, S. R. (1977). Predictive validity of student perception: A cross-cultural replication. *American Education Research Journal*, *14*(1), 45–49.

Yager, R. E., and McCormack, A. J. (1989). Assessing teaching/learning successes in multiple domains of science and science education. *Science Education*, *73*(1), 45–58.

Yager, R. E., and Roy, R. (1993). STS: Most pervasive and most radical of reform approaches to science education. In R. E. Yager (ed.), *What research says to the science teacher, Volume 7: The science, technology and society movement* (pp. 7–13). Washington, D.C.: National Science Teachers Association.

Yager, R. E., and Tamir, P. (1992). Accomplishing STS instruction and the student learning that results. In M. O. Thirunarayanan (ed.), *Handbook of science, technology and society, Volume I: A theoretical and conceptual overview of science, technology and society education*. Tempe, Ariz.: Arizona State University, College of Education.

CHAPTER 8

CREATIVITY AND THE VALUE OF QUESTIONS IN STS

John E. Penick

CREATIVITY AND SCIENCE

As Brandt (1986) pointed out, creativity is a personal way of using and directing one's own abilities. In the process, the creative person may restructure the problem rather than merely seeking solutions to the problem presented. Questions about the problem needing solution often, then, become questioning of the nature and even existence of the problem itself. Such questions give rise to ideas that would never have been considered initially.

Beginning with Socrates, scholars have pointed out that questions are the midwives that bring ideas to birth. Many great thinkers, scientists, and inventors gave credit for their inspirations to rather unscientific ideas and thinking. And, as Einstein put it, "The formulation of a problem is often more essential than its solution, which may be merely a matter of mathematical or experimental skill" (Getzels, 1975, p. 12). Einstein went on to note that imagination was more important than knowledge, that knowledge only grows when the mind is receptive to the unfamiliar and when old things are perceived in new ways. "To raise new questions, new possibilities, to regard old questions from a new angle, requires imagination and marks real advance in science" (Getzels, 1975, p. 12). A creative solution is the response to a creative question. Until a question is posed, no problem exists to be solved. Smilansky and Naftali (1986) provide data to support the notion that creativity is, in part, the ability to pose high-level problems and questions. Gertrude Stein is reported to have said, "If no one asked the question, what would the answer be?" Problem and question finding is a truly creative skill (Soriau, 1881) that must be cultivated. Wakefield (1992) and others have long noted that creativity is called for in identifying problems and that well-phrased questions are often the heart of creative investigation.

Dillon (1982) pointed out that few abilities or accomplishments have been praised or rewarded more than problem finding. Yet, compared to the volume of literature on problem solving, there is almost nothing written about the process or learning of problem identification. As a result, no theory of problem identification has been put forth. Dillon goes on to note that problem finding, including discovering, formulating, posing, may represent a more distinct and creative act than finding a solution. Many writers (Getzels and Csikszentmihalyi, 1975; Mackworth, 1965) conclude that question posing and problem finding are crucial, at the heart of originality, and form an extremely strong association with creativity. Yet, in most educational endeavors, problem finding is ignored to concentrate on the more mundane aspects of solving problems presented by the text, the teacher, or worksheets.

Not only is creativity an integral part of science, creative persons have been shown to be more observant (Barron, 1963); have better self-concepts, more self-confidence (Davis, 1983), and self-reliance (Moravcsik, 1981); more willing to take intellectual risks (Rubin, 1963); and able to score higher on normed achievement tests (Getzels and Jackson, 1963) as well as other achievement measures (Blosser, 1985). Creativity also affects gender roles as well with creative men more willing to accept traditional feminine interests (Bem, 1974), while creative women are seen as being more assertive and taking more risks. Creativity is sometimes called an androgynous trait, one that combines the best of masculine and feminine roles, ideas, and actions (Helson, 1967).

Creative people also desire to evaluate their own work, a pattern in those with strong self-confidence and inner direction (Dacey, 1989). Self-evaluation reflects the risk-taking and adventurous nature of many creative people as well their regular reliance on intuition and reflection as essential parts of the creative process. Besides, since creativity itself is often the person's goal (Brandt, 1986), who but the individual is in the best position to provide a creative evaluation?

Obviously, we want and need creative students in our classes and creative citizens in our future. Unfortunately, being creative often implies being unconventional (Moravcsik, 1981). Far more unfortunate, creative students, often the ones with the greatest potential, are overrepresented among high school and college dropouts (Davis, 1983). On a more positive note, many studies indicate that creative thinking skills can be learned with practice (Cronin, 1989) and creativity can lead to higher achievement scores (Getzels and Jackson, 1963). How students learn and their roles in the classroom determine to a large extent how much opportunity for creative learning they actually have. The role of the students must be carefully considered and planned for in a creative learning environment.

The Role of the Student

In general, students prefer to learn in creative ways by exploring, manipulating, testing, questioning, experimenting, and testing ideas (Torrance, 1963). Individuals are naturally curious and their curiosity and creativity "are stimulated by relevant, authentic learning tasks of optimal difficulty and novelty for each student" (APA, 1993, p. 7). In doing such tasks, if they are appropriate in stimulus and difficulty, we have many questions about what we are doing, what will happen if . . . ? and questions about causes and consequences. All of the questions are an integral component of both science and creativity. And, equally important, doing the task allows us to generate our own questions rather than merely relying on the questions provided by others.

Questions one raises independently are often the most effective way to ensure relevance to the individual. Certainly, we are more motivated to answer questions if we have raised them ourselves rather than ponder the questions of others. This is especially true in classrooms where the questions are often found in the text that tend to be stale, old, and boringly impersonal and generic.

Through their own questions, students create clear mental images of objects, phenomena, and their own understanding. Questions further serve to delineate problems, potential solutions, and other points of view. Questions are often a way of playing with ideas. A close relationship exists between play, creativity, and developing strategies for problem identification and resolution (Iverson, 1983). Such playing with ideas occurs when restraints on time, materials, and tasks are reduced and generated ideas are treated with respect. And, like most play, self-initiated action is the most satisfying to the learner.

Students with creativity, curiosity, and questions often desire to communicate (Risi, 1982). When one discovers, does, or invents something, a natural first response is to let others in on the excitement. This, "Hey, Mom! Look at me!" syndrome is a feature of both creativity and of science. Without communication of ideas, science would not exist as we know it. Chaudhari noted that, "Students' questions are their curiosity in action, their mind hunger" (Chaudhari, 1986, p. 31). But, if students are to communicate effectively and to formulate and follow up on questions, they must have a classroom climate conducive to their development. The teacher plays a key role in creating that environment where creativity is valued, encouraged, modeled, and rewarded. This environment is well exemplified by the STS classroom.

A variety of studies have examined creativity as a result of STS instruction. Myers (1988), Yager and Ajam (1991), and Yager and Ajeyalemi (1991) used the Torrance Tests of Creative Thinking. In all cases the investigators found that students scored significantly higher after experiencing STS classes than after learning in a more traditional classroom.

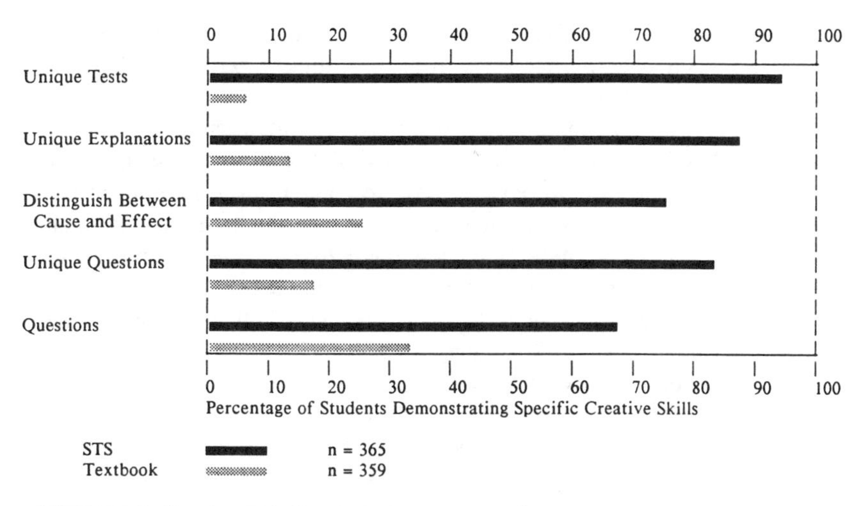

FIGURE 8.1 Creative Thinking in STS and Textbook Courses

Mackinnu (1991) in a similar study using fifteen creativity measures found that STS students showed significantly more gain on every item than did students from more text-oriented classrooms.

These results are what one would expect considering that the STS classroom encourages student ideas, initiative, and communication with other students. These results are also significant in the sense that, as teachers, we all want students who can raise questions, suggest causes, and predict consequences. Yet, while all STS teachers are overtly seeking these outcomes, more typical teachers only hope for them as a by-product of didactic instruction.

In addition to all these studies having similar findings, there is a common thread that ran through all the STS classes that produced such significant results. This thread was a stimulating classroom climate where students' questions and ideas were valued, their initiative encouraged, and where evaluation was based on a wide variety of criteria. This classroom climate, an essential element for both creativity and STS teaching, is made possible only by the teacher.

The Role of the Teacher

Creativity does not happen by chance alone. Teachers wishing to teach for creativity must carefully and consistently structure the classroom to maximize opportunities for creative endeavor, paying special attention to time, teacher role, available materials, and expectations of and for students.

Teachers must provide students with considerable intellectual freedom and safe opportunity and time to be spontaneous, explore, test, decide courses of action, and take risks. Students will not ask questions if they feel they and their questions may be pushed aside, rushed, or subject to ridicule. A rush to judgment is the opposite of creativity.

In creating a safe environment, one must treat all questions as valid and important. As soon as a teacher labels a question trivial or even tangential or divergent, some students will vow never to respond again. And, in reality, no questions are trivial if they are approached as evidence of what the student thinks is important or needs to know. While seemingly trivial, a question may give a teacher insight into the student's thought process, logic, or experience. After a simple question, ask the student to clarify or elaborate and often one will find they answer their own question. Routinely, teachers find that students can answer their own questions but lack confidence in their answers or want teacher approval.

Interestingly, providing students with creative and safe time enhances creativity regardless of the role of the teacher (Carter and Torrance, 1978). A teacher who systematically designs the classroom for development of questions and creativity is even more likely to have desired results. We know well that students tend to copy the teacher's behavior (Anderson and Brewer, 1946). Since much of creative thinking centers around experience, observations, questions, and possibilities, so, too, must the teacher's behavior.

The teacher is quite important in enhancing such creativity. Well-posed questions stimulate thinking, revealing alternate points of view and logic, and may be viewed as the embodiment of curiosity. But, to be a model of creative inquiry, a teacher must use questions that go beyond mere description. Questions to stimulate creativity must require and allow multiple possible answers and demand action. Questions model thinking as relevant problems are pursued. Questions act as windows on the phenomenon in question as we ask a question, select the best answer, and ask the next logical question, continuing the process until the desired evidence or explanation has been revealed.

Often, the tendency is to ask the ultimate question, "Why?" When a phenomenon is introduced and teachers ask, "Why did that happen?" students are put off because the "why" sounds very absolute and threatening. "Why" implies someone knows (or should know) the answer or is even possibly wrong ("Why did you hit your little sister?"). A better approach is to begin with the concrete, asking questions about what students did or observed. Then, ask how they might do it differently and what might happen if . . . ? Predictions are a reasonable next step as well as questions seeking to determine relationships with other, similar phenomena. Since we are consciously modeling good question asking behavior, these types of questions follow a logical hierarchy that student can emulate. We want them to delve into the problems and these questions assist in that endeavor. The "why" questions sound like test questions and are best if never asked. Table 8.1 suggests a simple hierarchy of questions to help organize the questions you ask.

TABLE 8.1 A Hierarchy of Questions

1. *Ask questions that describe—*
   What did you do?
   What happened?
   What did you observe?
2. *Ask questions that predict—*
   What will you do next?
   What will happen if you . . . ?
   What could you do to prevent that?
3. *Ask questions that relate to other phenomena—*
   How does that compare to . . . ?
   What did other people find?
4. *Seek explanations—*
   How would you explain that?
   What causes it to happen?
5. *Ask for evidence—*
   What evidence do you have for that?
   What leads you to believe that?

We should ask questions to obtain information, not to test students. When we seek information, we do not ask questions if we already know the answer. In good adult conversation, adults ask each other questions to find out, not to examine. We would not spend much time with an adult who continually quizzed us, particularly if they followed up by evaluating our answers. Our students are not different except they are captives of our classroom. As a rule of thumb in the classroom, if you wish to stimulate student involvement and creativity, never ask if you already know. We should also seek opinions and points of view such as, "How would *you* design an experiment to . . . ?"

To stimulate multiple answers, we must accept all answers, regardless of how good they may be. To encourage students to tell us their thinking, we must show them that each of their ideas has value, that we are paying attention to them. And, since evaluation stifles creative thought and reduces thinking to imitation, we must avoid judgment. But this does not mean we let everything pass by without comment nor is evaluation avoided.

Evaluation, rather than arising from the teacher's singular perspective, the norm, or tradition, must be based on causes and consequences. For instance, a student, using a two-pan balance asks, "Is this balanced?" The teacher could easily say "Yes" or "No" and get on with the activity. But such a response does little to help the student understand or be confident. Think how different is the scenario where the teacher, instead, replies, "What do you mean by 'balanced'?" The student is now forced into considering the meaning and the teacher learns more of the student's thought process. And, in talking it through, the student will probably answer his or her own question. In real science, evaluation always comes from causes and consequences, not from authority alone. Why should our science classes be any different?

Students may also become dependent on the teacher for the praise associated with positive evaluation. And, the more dependent students become, the less likely they are to stray from the conventional and the teacher's certain praise. All scholars and practitioners who study creative thinking agree that evaluation must be totally avoided in the early creative stages and, when evaluation does appear, must arise from individuals as they analyze their ideas, actions, and products in terms of personal needs, utility, prediction, or logic.

In the Classroom

Teachers can easily make a difference by the structure they provide and the atmosphere they create. If you wish to see creativity flourish in your classroom:

Provide opportunities for creative work;
(Time, materials, expectations)
Ask questions that demand answers;
(no "yes/no," recall, or answers you already know)
Wait for responses;
(Don't rush, if you really ask a question, wait for the answer. And wait again for multiple responses)
Accept unusual ideas, questions, or products;
(No judgment, just acknowledge and ask for more)
Ask students to examine causes and consequences;
(If that's true, then . . . ? , What may have caused that?)
Allow students to make decisions;
(Structure activities so that decisions must be made and allow students to do so)
Model creative thinking, action, and decision-making.
(Ask questions yourself, express curiosity, make the classroom stimulating.)

As teachers model creative behavior, they ask good questions, wait, expect, and encourage multiple responses, and seek evaluation based on causes and consequences. Teachers and students change. The change will be in the direction of being more creative, having more ideas, seeing more possibilities. To see more possibilities, students must also become more observant.

TEACHERS HELPING STUDENTS

Students need something to view if they are to make observations and raise questions and we must teach them how to observe carefully and systematically. Typically, students are told to observe, observations are counted and commented on, and the lesson moves on. Instead, if a teacher's goal is better and more creative observations, he or she must follow bare observations with teaching strategies designed to help students learn to observe more and in new ways and to see questions and possibilities as they work. If students exhaust their supply of simple observations, they should be helped to focus on *applications* (the candle burns string, will burn hair, won't burn foil, . . .), *possession* (her candle is burning, Megan's candle is burning, Lucas's candle is not burning, . . .), *location* (the candle burns on the table, sideways, while swinging, . . .), or *time* (the candle is burning now, is burning at 9:07, is still burning at 10:00, . . .). Each of these observations could also be used to generate questions about the phenomenon being observed.

These are not trivial observations. Each, in a scientific context, could potentially carry considerable meaning. But students who do think of these observation strategies may feel they are not worthy because they are unconventional and few students will think along these lines. Increased creativity, however, allows students to stray from the norm, to take risks, to seek the unusual. The creative individual looks at one thing and sees another; asks questions with difficult answers, and sees value in unusual ways.

Research by Carlsen (1990) revealed that teachers usually ask far more questions than students. While questions are usually viewed as positive, Dillon (1978) pointed out that teacher questions, rather than being stimulants to student thought, are often depressants. How can students think or work if they are bombarded with questions? How can they learn to structure their own questions if the teacher is the one who does the structuring? Fortunately, it has been noted (Carlsen, 1990) that there are more questions in a laboratory setting than in more traditional instruction and the ratio of student questions to teacher questions goes up quite dramatically. In addition, student rates of questioning went up in small groups. This research provides strong evidence that, if student questions are a serious goal, we can structure the classroom environment to maximize the generations of those questions.

Teachers who ask extended thought questions, listening and acknowledging student ideas, and observing are more likely to encourage creativity than such behaviors as giving directions, transmitting information, and asking recall questions (Foster and Penick, 1985). And teachers must help students know their ideas have value for their own sake, as stimuli for others, and in varied applications. Table 8.2 provides some practical suggestions on how to create an environment where questions work best.

TABLE 8.2 Some Practical Tips on Asking Questions and Creating a Safe Environment

1. Ask questions when you want information; don't ask test questions.
2. Seek opinion, perception, application, or experience rather than fact or accepted standard.
3. Do not rephrase questions immediately; if you are unclear, students will let you know.
4. Wait, sometimes interminably, after asking any question. Wait again after a response.
5. Create a safe environment by accepting, without evaluation, responses to your questions. Allow students to evaluate their own ideas.
6. Use student responses in your next question or statement.
7. Ask students to clarify or elaborate on their answers; never elaborate or clarify for them.
8. Think of your questions with students as an adult conversation; elicit ideas, use those ideas, and make it a two-way conversation.
9. Provide stimulating opportunities and materials for creative work.
10. Make the environment safe for exploring, risk-taking, experimentation, and speculation.

ENHANCING CREATIVITY

What is interesting about enhancing creativity is that usually one needs not make creativity the dominant issue. Rather, as changes in teaching behavior are made creative, growth results by accepting and encouraging divergent thinking, delaying judgment, and using stimulating brainstorming activities (McComas, 1989). Unfortunately, one of the most difficult tasks is to ask more open-ended questions. Being less directive, a behavior that is necessary to stimulate both thinking and good questions requires conscious restraint on the part of the teacher as well as a well-thought-out rationale for teaching. Teachers will not ask better questions or create an environment conducive to learning just by telling themselves they will. One must plan, practice, and reflect on one's own teaching. And, even then, it takes considerable time and effort before one achieves an acceptable level.

But the results are worth it. Consistently, the best teachers are described as encouraging discussion and conversation by accepting what students have to say and think. Teachers will change as their behaviors change. Effective STS teaching demands it. At the same time STS teachers are viewed differently, as persons who both care as well as being capable and competent.

REFERENCES

American Psychological Association (1993). *Learner-centered psychological principles: Guidelines for school redesign and reform.* Washington, D.C.: American Psychological Association and Mid-Continent Regional Educational Laboratory.

Amibile, T. M. (1989). *Growing up creative.* New York: Crown Publishers.

Anderson, H. H., and Brewer, J. E. (1946). Studies of classroom personalities, II: Effects of teacher's dominative and integrative contacts on children's classroom behavior. *Applied Psychology Monographs, No. 8.*

Barron, F. (1963). The need for order and disorder as motives in creative activity. In C. W. Taylor and F. Barron (eds.), *Science creativity: Its recognition and development* (pp. 153–162). New York: John Wiley and Sons.

Bem, S. L. (1974). The measurement of psychological androgyny. *Journal of Consulting and Clinical Psychology, 21*(3), 194–210.

Blosser, P. E. (1985). Investigations in science education. *Investigations in Science Education, 11*(3), 123–131.

Brandt, R. S. (1986). On creativity and thinking skills: A conversation with David Perkins. *Educational Leadership, 43,* 12–18.

Carlsen, W. S. (1990, April). Teacher knowledge and the language of science. Paper presented at the annual meeting of the National Association for Research in Science Teaching, New Orleans, La.

Carter, J., and Torrance, E. P. (1978). Abstract: Teacher effects in using creative thinking activities with sixth graders. *Journal of Creative Behavior, 12*(3), 217.

Chaudhari, U. S. (1986). Questioning and creative thinking: A research perspective. *The Journal of Creative Behavior, 20*(1), 30–33.

Cronin, L. L. (1989). Creativity in the science classroom. *The Science Teacher, 56*(2), 34–36.

Dacey, J. S. (1989). Discriminating characteristics of the families of high creative adolescents. *The Journal of Creative Behavior, 23*(4), 263–271.

Davis, G. A. (1983). *Creativity is forever.* Dubuque, Iowa: Kendall/Hunt.

Dillon, J. T. (1978). Using questions to depress student thought. *School Review, 36,* 50–63.

Dillon, J. T. (1982). Problem finding and solving. *Journal of Creative Behavior, 16*(2), 97–111.

Foster, G. W., and Penick, J. E. (1985). Creativity in a cooperative group setting. *Journal of Research in Science Teaching, 22*(1), 88–98.

Getzels, J. W. (1975). Problem finding and the inventiveness of solutions. *Journal of Creative Behavior*, *9*(1), 12–33.

Getzels, J. W., and Jackson, P. W. (1963). *Creativity and intelligence*. New York: John Wiley and Sons.

Getzels, J. W., and Csikszentmihalyi, M. (1975). From problem solving to problem finding. In I. A. Taylor and J. W. Getzels (eds.), *Perspectives in creativity* (pp. 221–246). Chicago: Aldine Publishers.

Hanks, K., and Parry, J. A. (1983). *Wake up your creative genius*. New York: William Kaufman.

Helson, R. (1967). Sex differences in creative style. *Journal of Personality*, *35*(2), 214–233.

Iverson, B. K. (1983). Play, creativity, and schools today. *Phi Delta Kappan*, *63*, 693–694.

Mackinnu. (1991). *Comparison of learning outcomes between classes taught with a science-technology-society (STS) approach and a textbook oriented approach*. Unpublished doctoral dissertation, University of Iowa, Iowa City.

Mackworth, N. H. (1965). Originality. *The American Psychologist*, *20*, 51–66.

McComas, W. F. (1989). Sparking creative thinking with S/T/S education: The results of the 1987–88 Chautauqua workshops. *Chautauqua Notes*, *4*(8), 1–2.

Moravcsik, M. J. (1981). Creativity in science education. *Science Education*, *65*, 221–227.

Risi, M. (1982). *Macroscole: A holistic approach to science teaching*. A discussion paper, D-82/2. Science Council of Canada, Ottawa.

Rubin, L. J. (1963). Creativity and the curriculum. *Phi Delta Kappan*, *44*, 438–440.

Smilansky, J., and Naftali, H. (1986). Inventors versus problem solvers: An empirical investigation. *Journal of Creative Behavior*, *20*(3), 183–201.

Soriau, P. (1881). *Theorie de l'invention*. Paris: Librairie Hachette.

Torrance, E. P. (1963). Toward the more humane education of gifted children. *Gifted Child Quarterly*, *7*, 135–145.

Wakefield, J. F. (1992). *Creative thinking: Problem solving skills and the arts orientation*. Norwood, N.J.: Ablex Publishing.

CHAPTER 9

USING WHAT HAS BEEN LEARNED: THE APPLICATION DOMAIN IN AN STS-CONSTRUCTIVIST SETTING

Gary F. Varrella

USING LEARNING BEYOND THE CLASSROOM

The acquisition and mastery of specific concepts and skills is no longer an acceptable end goal for science students. Our destination lay in a broader and more comprehensive domain that includes a rich variety of experiences in all of the sciences and the related fields of mathematics and technology (Rutherford and Ahlgren, 1990). *The Content Core* (NSTA, 1992), a support document for the national Scope, Sequence, and Coordination (SS&C) Project, challenges students to ask:

1. How do we know?
2. Why do we believe?
3. What does it mean?

The broad goal of literacy and the specific questions defined by the Iowa-SS&C effort demand that students be capable of doing something with what they have seemingly learned. Application of fundamental science concepts that can be directly related to real-world questions and situations is needed. Establishment of a frame-of-reference that can be easily related to the students' previous experiences is the appropriate starting point. The actual ability of students to apply fundamental concepts successfully to new situations can also provide an assessment of the depth of their understanding and an evaluation of the overall worth of the learning experience.

This type of experience is characterized by the need of the student to use skills characterizing the process domain and higher-order thinking skills and to apply such skills and science concepts in new situations. This use of skills and concepts is broadly categorized under the rubric, "Application Domain," in

Iowa-SS&C assessment efforts (Yager, Kellerman, Liu, Blunck, and Veronesi, 1993). Ideally, the students are responsible for their own learning experiences and the teacher serves as the facilitator of that process. The expectation that students become active in planning their own science learning experiences (within limitations of the wider course requirements and local expectations) makes vested interests out of each and every student—lending personal meaning and purpose to the topic of study. The necessity to think, to learn to relate science to the real-world, and to make decisions that are based on evidence rather than on speculation and peer influence, facilitates and promotes behaviors desirable of responsible members of an informed citizenry of a scientifically developed nation (Science Advisory Committee, 1985; Science Curriculum Framework and Criteria Committee, 1984; Wollman, 1983).

Iowa has been an active participant in the reform and restructuring movements sweeping primary and secondary science education. There are no time-tested answers to the questions surrounding restructuring of kindergarten through twelfth-grade science education; however, Iowa science educators have amassed a wealth of formative experience as well as results from related research.

This chapter will address four related topics. First, the rationale for the constructivist approach; second, consideration of the actual domain of application as interpreted by many Iowa science teachers; third, a brief review of four selected research reports related to the topic of application; and fourth, two situations will be introduced, one actual and one hypothetical, based on experiences of Iowa science teachers who are involved in the Iowa-SS&C Project. All discussion relates specifically to the central topic—the ability of students to *do something* with their acquired knowledge and understanding of science.

THE MILIEU OF APPLICATION

McCormack and Yager (1989) proposed a taxonomy for science education that included five domains (Yager et al., 1993).[1] This structure goes beyond the typical view that kindergarten through twelfth-grade science education should center only on the mastery of specific concepts and processes. Concepts and processes worth knowing must be usable and should be applied in a context that begins at a point of relevance (and necessarily at an existing level of understanding) for the learner. Success in accomplishing these tasks is measured in Iowa within the broad domain of application (Yager et al., 1993).

A potentially more familiar interpretation of "application" has been described in the taxonomy developed by Bloom (1956). The major categories of this taxonomy are listed below. They can be considered hierarchical in nature, as arranged here, ranging from least to most complex in terms of the learners' thinking processes:

Knowledge—remembering something learned or encountered previously;
Comprehension—understanding material without relation to anything else;
Application—solving a problem by using a general concept;
Analysis—breaking an idea or concept down into its parts;
Synthesis—using parts of learned ideas or concepts to make a new one; and
Evaluation—making a judgment about the worth or value of something.

A comparison of these two taxonomies is provided in Figure 9.1. Essentially, the applications domain of McCormack and Yager (1989) embraces *all of the* higher order thinking processes and experiences defined by Bloom (1956).

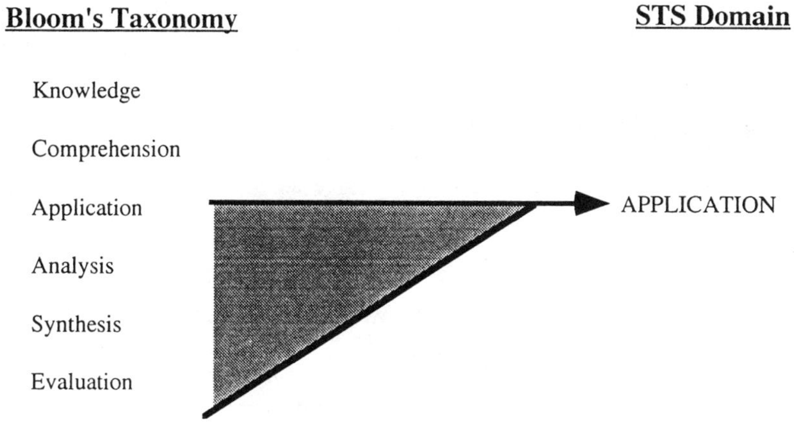

Bloom's Taxonomy **STS Domain**

Knowledge
Comprehension
Application → APPLICATION
Analysis
Synthesis
Evaluation

FIGURE 9.1 Application—Two Interpretations

SUCCESS IN "APPLICATIONS" IN IOWA: A BRIEF SUMMARY OF RECENT RESEARCH

Notable and consistent gains in students' abilities to use what they have learned in new situations have been documented in Iowa STS science classrooms (Mackinnu, 1991; McComas 1989; Myers 1988; Yager, 1990). Three of these studies (Myers, Yager, and Mackinnu) used a simple experimental design including both treatment and control groups. The fourth, by McComas (1989) was a case report.

The case report by McComas (1989) included a large number of students (1,289). The particular value of this report relates to the background and experiences of the teachers involved. All participants were in the Iowa Chautauqua Program which, for most teachers enrolled, is a first introduction to STS (total first-year experience: approximately seventy-five hours of inservice). From this first brief inservice experience for 65 percent of the students of these teachers to show a statistically significant (at $p < 0.05$) improvement of ability to apply the science they just learned is remarkable. All of the tests were teacher-generated (i.e., by Chautauqua participants), which invites inconsistency and a greater margin of error. However, all of the participants had been introduced to *The Iowa Assessment Handbook* (Yager et al., 1993), which is designed to assist novice STS-constructivist teachers in authentically assessing in the STS-constructivist environment. Further, ten years of experiences from the Iowa Chautauqua Program and three years of experiences from the Iowa-SS&C Project indicate a degree of consistency in teacher successes in authentically assessing student gains in the applications domain (Yager, Liu, and Blunck, 1993; Yager, Liu, and Varrella, 1993).

The Mackinnu (1991), Yager (1990), and Myers (1988) studies, on the other hand, dealt with smaller numbers of teachers and classes (a total of no more than twenty-nine different teachers among the three), but reported 100 percent improvement in the STS-constructivist classrooms. Students were able to *apply* science concepts to new or unique situations. A problem-solving approach was often the method of choice in the application situations. In certain instances, the learning situation grew out of previous STS experiences and extended into other related topics of science study.

These studies employed teachers who were both experienced and successful in teaching using the STS approach. The environment of each of the three studies was carefully managed and included control or "contrasting" groups of students experiencing more traditional modes of teaching. For this brief review, grouping these three studies is useful, but it must be remembered that pre- and posttesting procedures varied.

The relationship between teacher experience and student success has not been quantified using the STS *and* constructivist approach to teaching; however, a wealth of anecdotal data collected from Iowa kindergarten through twelfth-grade science teachers attests to the positive influence of the practice on eventual student successes in the Application Domain (Yager, Liu, and Blunck, 1993; Yager, Liu, and Varrella, 1993).

CLASSROOM EXPERIENCES

"What does an application situation look like in an Iowa classroom?" The discussion that follows is drawn directly from a module developed by two South Central Iowa Centers (SCIC)[2] where teachers work cooperatively on materials development based on actual middle school experiences.

The title of the module is Station E.A.R.T.H.[3] The unifying theme for the module is discovering and exploring things about our planet. An example used to stimulate thinking and encourage questions is built around Biosphere II, a project in Oracle, Arizona, which is a sophisticated attempt at duplicating Earth's unique conditions artificially (SCIC, 1992). Through student-centered exploration and applications to real-life (and often local) situations and relevant issues the students in the SCIC explored a series of science concepts at a variety of levels of complexity.

The relationships between the body of scientific knowledge and information and the uses of technology and the societal implications were all woven into the fabric of the experience. Included, for example, are activities related to food chains, the water cycle, the carbon and nitrogen cycle, and taxonomies. Issues related to conservation, recycling, human impact on the environment, and land use policy also have been explored by students in the classrooms of the SCIC. The learning environment in these classrooms is open and dynamic. There is ample opportunity for students' questions and input into the organization of the science experiences. The mutually agreed on (between teachers and students) experiences growing out of the module all revolved around the central theme of learning more about our planet.

The students in the classrooms of the Iowa-SS&C teachers are challenged to make sense of what is taught by trying to fit it with their experiences. Consequently, the words, concepts, the processes, and ideas that are explored and mastered through studies based on the Station E.A.R.T.H. module grow out of personal constructions of meaning and value. The result—eventual inclusion of new constructions—built on previously held knowledge and experiences, into the student's personal knowledge schema. The teacher is the facilitator in this setting; however, often the teacher also becomes a participant and learner, especially when a new related science study topic (extending beyond the current area of focus) emerges. Using a flexible approach to problem solving allows for adaptations, refinements, and the establishment of collateral lines of questions and studies as the students apply their existing and newly acquired knowledge and/or confront their misconceptions.

As with other Iowa modules, Station E.A.R.T.H. is open-ended, offering ideas, approaches, and core activities. No one specific procedural structure is described or implied. Rather, the recommendation is that the module be used as a catalyst for learning and study related to a topic (in this instance ecology, biomes, and the environment) for any science classroom. Consequently no two groups of students (and their teachers) are likely to approach the topic in an identical fashion. However, the eventual outcomes are surprisingly consistent in situations where the teachers have mastered the alternative (student-centered) paradigm of STS-constructivist teaching (Yager, Liu, and Varrella, 1993).

Two examples follow. One "trial success story" is drawn directly from the module; and the second example is a construction based on one of the "core activities" found in Station E.A.R.T.H. This second example also serves as an illustration of the effectiveness of an open-ended experience in addressing a series of predetermined outcomes.

This first example describes an actual experience by an SCIC (1992, p. 5) teacher and includes the early exploration and experiences related to the tinkering or messing-about stages of learning (using the Learning Cycle as a referent [Lawson, Abraham & Renner, 1989]) and the eventual challenge provided by a problem situation of the students' choosing:

Trial School Success Story

Students in the seventh and eighth grades at St. Malachys School in Creston, Iowa, took a journey to Station E.A.R.T.H. They began by formulating questions about the nine biomes introduced in the Station E.A.R.T.H. module. The students had brainstorming sessions about the essentials for survival and the needs of all living organisms. This was the first step in preparation for a proposed one year pilgrimage to a biome.

Students worked in cooperative groups of three or four individuals. They investigated topography, living and nonliving substances characteristic of their biome, and talked about how they would satisfy their own needs while living in the biome. The groups made use of a wide variety of resources, including the *Grolier's Encyclopedia* on the Mac Plus computer, and the videodisc player for a visual encounter of what the biome would look like.

The groups had to decide how much and what materials they would take with them on the trip. They also had to determine what materials would be available in the biome for food and shelter. They had to decide how to deal with the waste products they would create. Questions were generated continuously both within and between groups.

Students worked with enthusiasm preparing for a presentation before their "departure" to their biome. The presence of the true excitement of learning was particularly evident in the student-to-student interactions.

Core Activities Construction

The second example is based on an activity titled "Land Usage" (SCIC, 1992). This activity is one of the "core activities" included in Station E.A.R.T.H. The objective is the study the impact of a development (actual or fictitious) on the students' community. The students are expected to address environmental, energy usage, economic costs and benefits, and aesthetic considerations. The general approach is one of problem solving and the goal is to integrate all necessary aspects of science, technology, and society into "recommendations" that can be defended with substantive evidence. Figure 9.2 and Table 9.1 are representations of how students might go about dealing with the problems of land usage in their community.

The anticipated outcomes are listed in the center of each figure. On each side are representations of contrasting means of addressing those outcomes. A recent edition of *Heath Life Science* (Bierer, Lien, and Silberstein, 1987) was used as a model of a more objectivist approach. The questions on the right represent a more open ended (i.e., less rigid, more student-centered experiential approach to science teaching/STS-constructivist approach. The network of connections for learning and study between the questions and the targeted concepts are demonstrated by the intersecting arrows.

Table 9.1 provides more detail, relating student questions and activities to two of the three targeted outcomes.

The STS-constructivist approach emphasizes individual student growth beginning at the point of the student's level of understanding of the world and of science, beginning with their questions. The frame-of-reference is familiar to the students because the situation is (actually or artificially) based on a real place in their community. A variety of more typically accessible (e.g., textbook) and less typically used (e.g., community resource persons, field trips to the "site," and use of data and information available from national and state sponsored computer accessible data bases) resources are used. Cooperative learning methodology promotes a healthy exchange of ideas and a forum for trying out potential alternatives, compromises, and solutions.[4] The activity culminates with final student recommendations. The creation of the "final report" requires higher-order thinking and interpretive skills as well as encouraging application of written and oral communication skills.

More open-ended approaches employing the STS-constructivist paradigm can meet predetermined outcomes and facilitate mastery of concepts by using student questions to jointly plan (teachers and students) relevant experiences and applications. Through applications, the learner must interpret (in this instance) a number of observations and extend those observations into some type of logical plan (complexity of the plan, of course, related to developmental levels of the students involved) to accommodate the situation and its inherent problems. Though frequently addressing traditional and ubiquitous expectations and goals for science students, it is expected that experiences in the STS-constructivist classroom will likely be less traditional in nature. Addressing real questions such as those above makes it difficult, if not impossible, to indulge in the ordinary "march through the textbook" method of teaching science. In the Iowa STS-constructivist classroom the text serves as a primary

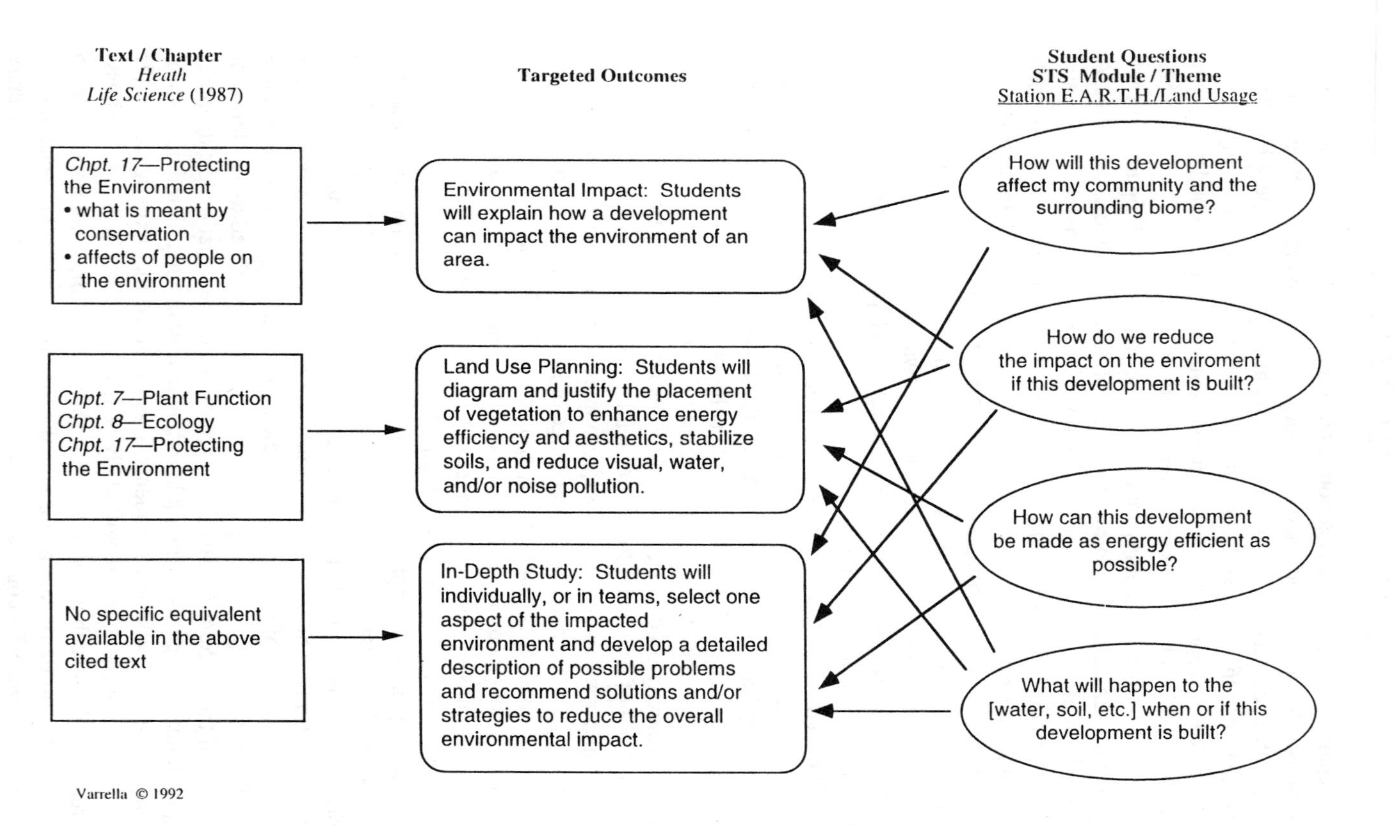

FIGURE 9.2 Comparison of Textbook and STS Approaches (Middle School Example)

TABLE 9.1 Comparison of Textbook and STS-Constructivist Approaches ("Land Usage" detail)

Textbook (Heath Life Science, *1987*)	*Targeted Outcomes*	*STS-Constructivist Approach*
1. Chapter 17—Protecting the Environment (example sections) a. Section (17.1) • what is meant by extinct and endangered • how people affect wildlife populations • how wildlife can be protected b. Section (17.3) • how the air, the water, and the land are polluted • what people are doing to protect the environment from pollution	Environmental Impact: Students will explain how a development will impact the environment of an area.	1. How will this development affect the environment of my community? *Activities:* a. Students will visit the site in teams and identify the components of the biome. b. Cooperating groups will identify, catalog, and describe existing environmental elements of the area. c. Students will identify community resource persons to assist them in identifying other potential impacts on the environment, the community, and other [impacted] surrounding areas.
2. Chapter 7—Plant Function Chapter 8—Ecology Chapter 17—Protecting the Environment (example sections) a. Section 7.3 • the functions of roots • how root systems differ b. Section 8.2 • what are the major biomes and their characteristics c. Section 17.4 • alternate sources of energy • what people are doing to conserve fossil fuels	Land Use Planning: Students will diagram & justify the placement of vegetation to enhance energy efficiency and aesthetics, and reduce visual, water, and/or noise polution.	2. How do we reduce the impact on the environment if this development must be built? *Activities:* a. Meet with city planners and discuss housing density, green belts, etc., and related developer costs. b. Identify the soil types; possible pollutants, and run-off; propose strategies to ameliorate such. c. Create a simple landscape design to enhance appearance and minimize erosion, noise, and visual pollution. d. Research the use of green plants as sources for breakdown of certain types of pollutants and toxics; propose and justify inclusion—if appropriate (and defend reasoning).

resource, but not as the central organizer. The most significant result of this practice is a distancing from the more typical objectivist approach to science teaching. Once the instruction is not externally driven by a highly structured map imposed by textbooks or other less flexible curricula, the degree and quality of interaction between students and teachers improves. The classroom environment remains a positive place where everyone has a stake in the plans for the learning experiences as well as the results. This classroom climate helps, as does the level of student enthusiasm, to keep interest high.

In the objectivist setting students often resolve their personal cognitive conflict by holding two divergent interpretations. One interpretation relates to the teacher, which focuses on his or her learning expectations in the classroom and a second interpretation for themselves, which relates to the students' personal schema based on their out-of-the-classroom experiences. In the STS-constructivist setting, the teacher is more likely to be sensitized to these preconceptions and to the mental processes by which the students make sense of phenomena. The teacher acts as the *facilitator* in the experience, which always allows for more teacher interactions with small groups and individuals—making it easier for the teacher to recognize and capitalize on teachable moments and promote individual growth. Also, the teacher can allocate the time necessary for the students to reflect on their experiences in relation to what they already knew or perceived and resolve discrepancies (deal with misconceptions) that arise (Lorsbach and Tobin, 1992).

When using a problem-solving approach, these discrepancies arise as the students apply their previously held beliefs and conceptions to specific challenges and the results and outcomes are not as expected. This is significant because until the personal cognitive conflict occurs, no resolution on the part of the individual learner is necessary. The concomitant requirement for successful student reinterpretation is a pedagogically able teacher who can grasp such "teachable moments" and facilitate resolution.

The enthusiasm of the students for science is greatly augmented as well because student interest in the [local] problem is high, as is ownership of the solution. Everyone, whether five or fifty-five, is more enthusiastic about a task when they have a greater *degree of ownership and choice in the matter* at hand.

LESS *IS* MORE

The learning experiences described in this chapter require the student to incorporate acquired, experientially based knowledge and comprehension to the solution of problems at Bloom's (1956) levels of application, analysis, synthesis, and evaluation. For the purposes of the STS-constructivist initiative all of the higher-order cognitive skills and the related experiences can be grouped under the general category or Domain of Applications.

"It is the quality of the science and technology education that students are actively engaged with that is important—not the quantity" (Kyle, 1984, p. 129). The STS-constructivist approach to teaching science may take longer and introduce and/or build on fewer concepts (i.e., less content is covered in a specified time frame) when it is compared with the traditional march through the textbook. This reflects the "less is more" philosophy—a conviction that is not confined to the STS-constructivist (and Iowa-SS&C) classroom alone. Other recognized reform movements such as Project 2061, the Coalition for Essential Schools, and the National Scope, Sequence, and Coordination Project echo similar themes (NSTA, 1992; Rutherford and Ahlgren, 1990; Sizer, 1992). Furthermore, the STS-constructivist approach may not be the sole reason that less material is covered. If the teacher's [and the community's] goals include "applications," then the number of concepts embraced in similar time frames will be reduced even without the added dimension of STS-constructivism. However, depth of understanding related to the topic is greater and there are many more opportunities and experiences requiring high-order thinking and learning abilities.

Sustained change in understandings and beliefs is a complicated and time-consuming prospect—both for the learner reaching a new cognitive equilibrium as a result of a specific experience(s) and for the teacher striving to establish and maintain an environment that is conducive to learning (Lorsbach and Tobin, 1992). Lyon and Gettinger noted in their study of student performance on knowledge, comprehension, and application that, "In general students need more time to attain mastery of school learning tasks as they advance up the hierarchy of [Bloom's] taxonomy" (Lyon and Gettinger, 1985, p. 18). They found that most (89%) seventh- and eighth-grade students in their sample could master knowledge tasks (i.e., correctly answer test questions) within a 5:1 time ratio, but only one-third (33%) reached mastery on [Bloom's] application tasks in a similar time. If it is agreed that students must be able to use what they learn to solve problems or construct solutions and explanations for previously unknown events and occurrences, the factor of time becomes less important and the efficacy of the approach becomes pivotal.

In science programs using an STS-constructivist approach, science is no longer a search for *the* truth. Rather, it becomes more like the science that scientists do—active, drawing on a variety of disciplines, and involving interactions within the social (and political) context. The broader goal for the student is the personal negotiation of a more accurate and refined understanding of science and its relationship to and impact on technology and society through classroom experiences. Through this process the learner can resolve discrepancies between what he or she knows and what seems to work in the new (problem) situation. The search process is necessary to work out a resolution to the problem (Andre, 1986) and will stimulate personal scrutiny of previously held beliefs and interpretations. If the alternatives considered by the learner provide a more effective resolution of the problem situation, then that individual will more likely incorporate it into his or her personal schema.

The learning partnership between the teacher and student which grows out of studies in the Domain of Application is truly an exciting and dynamic thing. The effective STS-constructivist teacher is the learning coordinator—facilitating and stimulating opportunities for learning, advocating a real-world frame-of-reference, and encouraging eventual cognitive restructuring in the mind of each student. The student must be involved firsthand with the trial and error process, accept his or her personal stake in the classroom learning environment, and individually work toward cognitive reequilibration as his or her preconceptions and misconceptions are challenged. As each student works through these mental building and restructuring processes, the learning that eventually results is more meaningful and enduring.

NOTES

1. Since this has been expanded to include six "domains"—concept, process, application [and connections], attitude, and (the newest addition) world view (Yager, Kellerman, Lui, Blunck, and Veronesi, 1993).

2. The SCIC includes the centers of the Chariton District Site and the Consortium of the six districts of Creston, East Union, Lenox, Mt. Ayr, Prescott, and St. Malachys—all of which work cooperatively on materials development.

3. For the curious: "E.A.R.T.H." stands for "Environmentally, Animals Rule Too Harshly!" This is part of a light-hearted introduction for the reader (assumed to be a teacher planning to use the module as a part of the course of science study).

4. Frequently, in such situations the teacher also becomes a partner in the learning experience. This type of participation also allows the teacher to model desirable learning habits and problem-solving approaches for their students.

REFERENCES

Bierer, L. K., Lien, V. F., and Silberstein, E. P. (1987). *Heath life science*. Lexington, Mass.: D. C. Heath and Company.

Bloom, B. S. (ed.). (1956). *Taxonomy of educational objectives, handbook I: Cognitive domain*. New York: David McKay.

Brotherton, T., Courtney, R., Eilers, C., Hoegh, J., Hoegh, S., Howie, F., Irelan, S., Jeffryes, C., Kimbel L., Kleinow, R., Lippincott, K., McCabe, M., Spargur, J., Wiele, L., Yossi, L., and Zumbach, G. (1992). *Station E.A.R.T.H. (Biomes and Environment)*. Iowa City: University of Iowa, Science Education Center.

Jakubowski, E. and Tobin, K. (1991). Teachers' personal epistemologies and classroom learning environments. In B. J. Fraser and H. J. Walberg (eds.), *Educational environments, evaluatory antecedents and consequences* (pp. 201–214). Elmsford, N.Y.: Pergamon Press.

Kyle, W. C., Jr. (1984). What implications for science education can be drawn from research on effective schools and classrooms? What school and home environmental factors influence student achievement and attitudes toward science? In D. Holdzkom and P. B. Lutz (eds.), *Research within reach: Science education* (pp. 123–141). Charleston, W.V.: Research and Development Interpretation Service, Appalachia Educational Laboratory.

Lawson, A. E., Abraham, M. R., and Renner, J. W. (1989). *A theory of instruction: Using the learning cycle to teach science concepts and skills*. Washington, D.C.: The National Association for Research in Science Teaching.

Lorsbach, A., and Tobin, K. (1992). Research matters—to the science teacher, constructivism as a referent for science teaching. *NARST Bulletin* (30).

Lyon, M. A., and Gettinger, M. (1985). Differences in student performance on knowledge, comprehension, and application tasks: Implications for school learning. *Journal of Educational Psychology, 77*(1), 12–19.

Mackinnu. (1991). *Comparison of learning outcomes between classes taught with a science-technology-society (STS) approach and a textbook oriented approach*. Unpublished doctoral dissertation, University of Iowa, Iowa City.

McComas, W. F. (1989). The application of scientific knowledge: The results of the 1987–88 Chautauqua workshops. *Chautauqua Notes, 4*(6), 1–2.

McCormack, A. J., and Yager, R. E. (1989). A new taxonomy of science education. *The Science Teacher, 56*(2), 47–48.

Myers, L. H. (1988). *Analysis of student outcomes in ninth grade physical science taught with a science/technology/society focus versus one taught with a textbook orientation*. Unpublished doctoral dissertation, University of Iowa, Iowa City.

National Science Teachers Association. (1990–1991) Science/technology/society: A new effort for providing appropriate science for all (Position Statement). In *NSTA handbook* (pp. 47–48). Washington, D.C.: Author.

National Science Teachers Association. (1992). *Scope, sequence, and coordination of secondary school science: Volume I. The content core. A guide for curriculum designers*. Washington, D.C.: Author.

O'Neil, J. (1992). Schools try problem-based approach. *The Wingspread Journal*, 1–3.

Rutherford, F. J., and Ahlgren, A. (1990). *Science for all Americans: A Project 2061 report on literacy goals in science, mathematics, and technology*. New York: Oxford University Press.

Science Advisory Committee. (1985). *Model curriculum standards, grades nine through twelve: Science* (1st ed.). Sacramento, Calif.: California State Department of Education.

Science Curriculum Framework and Criteria Committee. (1984). *Science framework addendum for California public schools: Kindergarten and grades one through twelve*. Sacramento, Calif.: California State Department of Education.

Sizer, T. R. (1992). *Horace's school*. New York: Houghton Mifflin.

Wollman, W. (1983). Models and procedures: A classroom study of teaching and transfer. *School Science and Mathematics*, *83*(2), 122–132.

Yager, R. E. (1990). Instructional outcomes change with STS. *Iowa Science Teachers Journal*, *27*(1), 2–20.

Yager, R. E., Kellerman, L., Liu, C., Blunck, S. M., and Veronesi, P. D. (eds.). (1993). *The Iowa assessment handbook*. Iowa City: University of Iowa, Science Education Center.

Yager, R. E., Liu, C., and Blunck, S. M. (1993). *The Iowa Chautauqua program annual assessment report 1992–93*. Iowa City: University of Iowa, Science Education Center.

Yager, R. E., Liu, C., and Varrella, G. F. (1993). *The Iowa scope, sequence, and coordination (SS&C) project assessment report 1990–93*. Iowa City: University of Iowa, Science Education Center.

Chapter 10

Enhancement of Opportunities for Low-Ability Students with STS

Eric Olson
Srini Iskandar

STS for Less Successful Students

A quote attributed to Immanuel Kant neatly sums up a constructivist notion of learning, "I Ought, Therefore I Can." Not only is this a succinct paraphrase of Rene Descartes famous "I Think, Therefore I Am," but it embodies the notion that all students regardless of their current level of development can perform any task set before them if they feel a sufficient desire to do so. Currently there is a mounting body of evidence that an interactive and student-centered curriculum adequately meets the special learning needs of low-ability students, and does so in a way that enhances learning and provides challenges for all students. Students of low ability function significantly better in the environment of an STS classroom.

Students who typically attain a grade of C or lower in science classes or who are disinterested and ill-motivated to achieve up to their capacity in a science classroom can be operationally defined as "low ability." This is not to say that these students intrinsically function on a lower level than other students; rather their abilities have not been sufficiently sparked by the subject matter or the strategies used by teachers. STS instruction provides a perfect vehicle for these students to make the science curriculum personally relevant and meaningful. The studies that are cited as evidence for these assertions provide evidence that STS instruction meets the needs of all students, but especially those most often not served in traditional schools and classrooms.

Several questions arise regarding the appropriateness of STS instruction for low ability students. How do these learners respond to the openness and student-centeredness of such instruction? How do these students respond to the use of appropriate technologies (computer, laser disks, interactive video, communications software)? Does establishing local relevance improve student-motivation? Does the shift in emphasis from content to context provide a fruitful basis on which learning can take place for low-ability students?

It is most often the case that students of low ability are those most in need of meeting success. The STS classroom provides the unique opportunity for these students to find success in science on their own terms. Usually it is the case that students of impaired ability are constantly reminded of their own inadequacies in the science classroom.

One can easily recall the typical scenario where a teacher prompts students to answer questions from their seats and only the few students who are attentive to the textbook-generated material can answer. It is worse when a teacher asks students who do not have an activity completed within an external time frame to raise their hands and a student of lower ability is forced to put the figurative dunce cap on. The shattering of self-esteem in such situations can almost be heard. Traditional classrooms are configured to remind students constantly of their inadequacies rather than emphasizing their strengths.

The latest incarnations of the national standards for science education are calling for "science for all Americans," including students of low ability. The STS movement is preeminently successful in striving and attaining this goal in meeting the educational needs of low-ability students. Several recent research projects regarding STS classrooms and STS activities have been completed affirming this point.

In a study "Student Achievement in Core Subjects of the School Curriculum" the following conclusion was advanced: "Another factor associated with high achievement was active learning. Students who said that their teachers required them to interpret and apply knowledge to the completion of tasks tended to score much higher on these assessments than did respondents who reported that their lessons were limited mostly to passive reception of knowledge through lectures and textbooks" (Patrick, 1991).

In attempting to characterize the "at risk" low achiever, Lehr and Harris (1988) use the following terms in an attempt to identify this group: disadvantaged, culturally deprived, underachiever, nonachiever, low ability, less able, low socioeconomic status, language impaired, dropout prone, alienated, marginal, disenfranchised, impoverished, underprivileged, low performing, and remedial.

In attempting to identify how these students are treated differently Lehr and Harris report that low-ability students are often:

1. Seated farther away from the teacher
2. Given less direct instruction
3. Offered fewer opportunities to learn new material
4. Asked to do less work

5. Called on less often
6. Given less wait time
7. Questioned primarily at the knowledge-comprehension levels
8. Not prompted when they do not know the answer to a question
9. Given less praise
10. Rewarded for inappropriate behavior
11. Criticized more frequently
12. Given less feedback
13. Interrupted more often
14. Given less eye contact and other nonverbal communication of attention and responsiveness.

It is no wonder given these parameters that Lehr and Harris conclude that these students tend to exhibit the following characteristics: academic difficulties, lack of structure (disorganized), inattentiveness, distractibility, short attention span, low self-esteem, health problems, excessive absenteeism, dependence, discipline problems, narrow range of interests, lack of social skills, inability to face pressure, fear of failure (feels threatened by learning), and lack of motivation.

The students referred to in this study are obviously troubled and are certainly in need of our professional attention.

In a recent study Iskandar (1991) compared STS and textbook-oriented classrooms for low ability students. The study concluded, "a significant difference (at .05 level) occurs for application of concepts, attitudes toward science, creativity skills (asking questions and predicting consequences), and science process skills for grades six, seven, eight, and nine low achieving students who were taught using the two methods." This study matched classes of students in terms of size, gender balance, socioeconomic level, ethnicity, and indicators of science interest. Controlling as much as possible for these variations is crucial for maintaining reliability of results. The Iowa Assessment package was used to test across five domains of science instruction (Concept, Application, Creativity, Attitude, and Process). These data were then collected and analyzed by t-tests and analysis of variance procedures to compare differences in pretest/posttest scores for low-ability students. In this carefully controlled study significant conclusions were drawn for the advantages that STS has for low-ability students.

In enhancing opportunities for low ability students STS does not abandon a rigorous curriculum. Yager et al. (1990) point out that although it is often feared that an STS focus would impair the acquisition of information, on the contrary it has been demonstrated that students retain as much information in an STS classroom as in a more traditional setting. Further, data drawn from the other domains (application, attitude, creativity, and process skills) lend support to the notion that the information that is retained is more meaningfully interwoven into a student's conceptual make-up.

This discussion would then lend credence to the argument that STS serves the dual purpose for low-ability students of finally providing the success that a traditional approach lacks while it also frames that experience in such a way as to be truly meaningful.

As has been pointed out in some of the earlier chapters in this monograph, the 1980s position statement of the National Science Teachers Association states:

> The goal of science education during the 1980s is to develop scientifically literate individuals who understand how science, technology, and society influence one another and who are able to use this knowledge in their everyday decision-making. Such individuals both appreciate the value of science and technology in society and understand their limitations. (NSTA, 1982, p. 1)

This position speaks clearly to the issue that science education must be made available to all Americans regardless of ability in this increasingly complex world. STS approaches and activities show promise in their ability to meet this goal.

In a recent study focusing on STS classrooms Liu (1992) found significant effects in STS instruction when contrasted with more traditional classes for students of low-ability levels in the following areas:

1. Low-ability students in classes taught with an STS approach learned more science concepts when compared to students in classes taught with a textbook-oriented approach.
2. Low-ability students taught with an STS approach developed science process skills significantly better compared to students in classes taught with a textbook-oriented approach.
3. Low-ability students in classes taught with an STS approach were able to apply more science concepts and principles when compared to students in classes taught with a textbook-oriented approach.
4. Low-ability students in classes taught with an STS approach were able to develop more science creativity skills when compared to students in classes taught with a textbook-oriented approach.
5. Low-ability students in classes taught with an STS approach developed more positive attitudes toward science when compared to students in classes taught with a textbook-oriented approach.
6. Low-ability students taught with an STS approach improved their understanding of the nature of science when compared to students in classes taught with a textbook-oriented approach.

7. Low-ability students in classes taught with an STS approach possessed more accurate perceptions concerning science careers than did students in classes taught with a textbook-oriented approach.

These conclusions are especially pertinent when considering the typical plight of low ability students in traditional classrooms. In light of the fact that students of low ability by definition are those students who experience the lowest level of success in traditional classrooms, it is not surprising that STS classrooms are most effective in enhancing educational opportunities for these students, and in attaining the goal of "science for all Americans" set out by NSTA and AAAS.

Educators must be especially concerned with those students who need the interactive and context-driven instruction that STS engenders. These are the students who benefit most from STS instruction. Research evidence shows that teachers can take advantage of an STS approach to education in facilitating growth in all students. However, STS engenders the much needed growth among low-ability students while also maintaining a significant level of challenge for students of higher ability.

Several studies have corroborated the point that students of low ability profit from an STS experience. Mackinnu (1991) collected data that affirm the above statements that students of low ability function better in an STS than a textbook approach in accumulating process skills, growing in the application of science concepts, expressing themselves creatively, and in maintaining a positive attitude toward science. This study was specifically designed to investigate the impact of STS instruction. It followed the progress of fifteen experienced STS teachers through the course of instruction. The *Iowa Assessment Handbook* was then used to measure growth in five domains (concepts, processes, applications, creativity, and attitude). The reliability of these instrument is in the 0.80 to 0.90 range for all of these domains. Appropriate t-tests were applied to pre- and posttests to establish levels of significance. In summarizing his findings Mackinnu concludes that the STS approach does provide significant advantage over the traditional textbook-oriented approach for both low- and high-ability students.

STS instruction is a significant departure from a traditional approach to teaching. These differences are crucial to the success of low-ability students. Students in this group blossom when they can take a more active role in their own learning. STS instruction shifts the burden of responsibility for learning from a teacher-centered to a student-centered approach, which by definition attains the goal of an individual-based lesson plan. In so doing students are enabled to meet success on their own terms and at their own pace. The focus of teacher as coach and facilitator as opposed to taskmaster and disciplinarian facilitates a positive and nurturing environment in which learning can take place. The focus on taking instruction out of the classroom and applying it in a real-world setting reinforces the fact that learning is a lifelong process and serves to further the self-reliance and conscious responsibility that must be an essential aspect of every citizen of the next millennium.

Myers (1988) has discussed student performance in the STS classroom. His findings indicate that there is no significant difference between student success with STS and traditional approaches in learning science concepts and in the uses of the science process skills as measured by the Iowa Test of Educational Development (ITED). The STS approach does have a significant impact on students' abilities to apply their science knowledge, to think creatively, and in improving their interest in science careers. It is indeed significant that STS students achieved at a similar level as did their textbook-educated peers in areas where STS instruction is not a focus and achieved significantly better in those areas where STS instruction concentrates. It is interesting to speculate on how low-ability students might fare on such "high stakes" tests as the ITED or SAT as developers prepare better instruments to test across all the domains of science.

Spade et al. (1985) in an analysis of characteristics contributing to an effective school in the area of mathematics and science suggest that the school setting significantly influences the educational experience of low-ability students. While this study did not emanate from an STS classroom, there are implications for analyzing the appropriateness of STS instruction for low-ability students. The results indicate that students of low ability are particularly sensitive to the atmosphere of the school. STS provides a unique opportunity to facilitate a milieu in which low-ability students can achieve. Spade concluded that STS was particularly important to the curricular organization created by the school and the academic encouragement given to the students. These are areas in which STS has been shown to excel.

It was Brown's view (1988) in "Effects of Self-Efficacy—Aptitude Incongruence on Career Behavior" that low-ability students believing in their ability to compete successfully with a variety of science and engineering majors facilitated their academic performance and persistence. Further, this study concluded that engendering self-efficacy had a significant impact on raising the grade-point average of low-aptitude students. Self-efficacy as defined by the ability to produce an effect is a notion that is integral to the STS approach. Students in the STS classroom concentrate on taking actions and monitoring results. It is an essential feature of a constructivist notion of science education that students be empowered by identifying the fruits of their own labor. Students who can use their education to take real actions seem to function in an increased capacity; this is especially relevant to students of low ability who may not have intrinsic motivation.

One of the fundamental aspects of STS instruction is the use of relevant technology in the science classroom. One of the obvious questions that arises here is: How does this use of technology effect low-ability students? In a 1989 study completed by the Texas Learning Technology Group entitled "Physical Science: Discovery through Interactive Technology" several conclusions were drawn, including:

1. Teachers reported that the technology worked best for students with learning problems.
2. The greatest differences were noted between low-ability students instructed using the technology and those students in a control group.
3. Low-verbal and low-quantitative students in classes where the use of technology was applied significantly outperformed students in a more traditional setting.
4. Students who where exposed to technology expressed a more positive attitude toward taking additional science classes than did students in more traditional classes.

These conclusions again validate STS instruction for low-ability students. The use of technology, though sometimes considered only within the capabilities of higher-functioning students, can in fact be used to enhance the education of all students, and is especially helpful in motivating low-ability students.

In a study conducted by Allen and Dietrich (1991) entitled "Student Differences in Attribution and Motivation toward the Study of High School Regents Earth Science" the authors summarized their work to define an appropriate science curriculum as lending support to encourage more students to take more demanding and higher-level science courses. Supportive methods should be used to provide opportunities for success that are clearly related to the effort put forth in mastering science concepts. Students ought to have the opportunity to see that their efforts can have a "pay off," then they may make an attribution shift that is focused on their efforts rather than their perceived lack of ability. STS is particularly important for students who are traditionally viewed as "low achievers."

STS instruction speaks directly to the challenges described in this study. The STS classroom itself functions as a support unit for all learners, drawing on diverse experience, and using those many facets as a resource for growth rather than a hindrance to instruction. The goals of teaching in the STS classroom are such that students are constantly recognizing the payoff of their efforts. All students, but especially low-ability students, can then initiate the attribution shift that will lead to greater success both in and out of the science classroom.

Though this study was conducted in a non-STS setting, the conclusions that they draw are clearly pertinent to the argument that STS is most appropriate for low-ability students. Students in STS classrooms focus their efforts on personal relevance and on taking action. Low-ability students in STS classes are then much more likely to be able to make the "attribution shift" from focusing on lack of ability to fruitfulness of effort.

Pearson (1991) in his study "Testing the Ecological Validity of Teacher-Provided versus Student-Generated Post-questions in Reading College Science Text" concluded that students generating and answering their own questions favorably impacts midrange (weekly quiz) performance. Student-generated questions based on quiz-related study reading had a tendency to induce prose processing of material relative to those dependent measures to a greater degree than teacher-provided questions based on the same text. These conclusions drawn from a decidedly non-STS setting lends credence to the effectiveness of the use of student-generated questions, an essential feature of the STS classroom.

In Kirkwood and Carr's (1989) "A Valuable Teaching Approach: Some Insights from LISP (Energy)" a study was undertaken involving a student-centered approach to learning the physical concepts of energy. Though not called STS, this program has much in common with STS instruction. The authors reasoned that approaches to learning that involve greater participation by the learners in setting the learning context and the learning agenda are typically seen to be relevant to classes of well-motivated students. These approaches are often overlooked for "low-ability" science classes where the learners are assumed to bring nothing to the classroom and to be generally incapable of student-centered learning activities. Teaching students who have been "turned off" to school and particularly science, can be a very dispiriting experience for teachers. Common coping strategies with low ability students include using a highly structured transmission mode of teaching in which material is assembled by the teacher, displayed to the learners, copied by them into books, and learned (or more commonly not learned) for examinations. Teachers control the learning totally, and in this way it is hoped that chaos will be avoided. This procedure traps teachers into expending a great deal of effort in telling students what they want them to know, based on what they do not know already. This study indicates that a student-centered curriculum can be as effective and as efficient with low-ability as with high-ability students.

Research studies have indicated considerable support to the notion that STS serves low-ability students well. This method of instruction focuses on local relevance, problem solutions, and student interaction. It draws in students who would otherwise be disenfranchised and disengaged.

One of the primary goals of STS instruction is a student-centered curriculum. The incredulous teacher who feels this is a wonderful goal but it just will not work in his or her classroom can take heart. The studies cited here lend credence to the notion that the STS philosophy can work in all classrooms, and is especially beneficial to students of low ability. These are the students who are most in need of our attention as educators. Past reform movements have really focused on fostering academic growth among the highest echelon of students. The mechanism for change that STS proposes is especially appropriate in this era of national standards calling for science education for *all Americans*. Low-ability students are most responsive to the interactive and student-centeredness of the STS classroom; they need the reinforcement and support that STS provides in order to meet with a growing level of success in their academic careers.

A typical junior- or senior-level science teacher in the United States probably does not have many low-ability students in his or her classes. Low-ability students are not required to take advanced science courses and have probably long ago been so "turned-off" by science that they no longer enroll in further courses. Students of low ability have been so bludgeoned by their previous courses that they identify science with personal failure and with unattainable concepts. These students who are disenfranchised from scientific learning may well form the majority of our citizenry in the future. If we are to stem the hemorrhage of talent out of the sciences, we must repair this wound.

The students who have been so blithely referred to as "low ability" in this chapter may not truly be deserving of such a stigma. In fact, they may simply be "low ability" because of the system in place to reinforce that idea. They may in fact represent exactly the sort of divergent and creative thinkers that we need as active participants in this increasingly complicated and technologically oriented world. All students can be participants in the learning process. Students who function poorly may do so because traditional modes of science education do not provide the means for them to meet with success. STS instruction does provide the interactive and context-driven instruction that all students can use to challenge themselves and grow regardless of ability.

REFERENCES

Allen, J. D., and Dietrich, A. (1991). *Student differences in attribution and motivation toward the study of high school regents earth science*. Unpublished manuscript.

Brown, S. D., Lent, R. W., and Larking, K. C. (1988). *Effects of self-efficacy-aptitude incongruence on career behavior*. Unpublished manuscript.

Iskandar, S. M. (1991). *An evaluation of the science-technology-society approach to science teaching*. Unpublished doctoral dissertation, University of Iowa, Iowa City.

Kirkwood, V., and Carr, M. (1989). A valuable teaching approach: Some insights from LISP (energy). *Physics Education*, *24*(6), 332–334.

Lehr, J. B., and Harris, H. W. (1988). *At risk, low-achieving students in the classroom. Analysis and action series*. Washington, D.C.: National Education Association.

Liu, C. (1992). *Evaluating the effectiveness of an inservice teacher education program: The Iowa Chautauqua program*. Unpublished doctoral dissertation, University of Iowa, Iowa City.

Mackinnu. (1991). *Comparison of learning outcomes between classes taught with a science-technology-society (STS) approach and a textbook oriented approach*. Unpublished doctoral dissertation, University of Iowa, Iowa City.

Myers, L. H. (1988). *Analysis of student outcomes in ninth grade physical science taught with a science/technology/society focus versus one taught with a textbook orientation*. Unpublished doctoral dissertation, University of Iowa, Iowa City.

National Science Teachers Association. (1982). *Science-Technology-Society: Science education for the 1980s*. Position Paper. Washington, D.C.: Author.

Patrick, J. J. (1991). Student achievement in core subjects of the schools curriculum. *ERIC Digest*. Columbus, Ohio: ERIC Clearinghouse for Social Studies/Social Science Education.

Pearson, J. A. (1991). Testing the ecological validity of teacher-provided versus student-generated post-questions in reading college science text. *Journal of Research in Science Teaching*, *28*(6), 485–505.

Spade, J. Z. (1985). *Effective schools: characteristics of schools which predict mathematics and science performance*. Unpublished manuscript.

Texas Learning Technology Group. (1989). *Physical science: Discovery through interactive technology. Status and evaluation*. Unpublished manuscript.

Yager, R. E., Blunck, S. B., Binadja, A., McComas, W. F., and Penick, J. E. (1988). *Assessing impact of S/T/S instruction in 4–9 science in five domains*. (ERIC Document Reproduction Service No. ED 292641.)

CHAPTER 11

BREAKING THE "MOLD"—STS ALLOWS CELEBRATING INDIVIDUAL DIFFERENCES

Janice Koch
Susan M. Blunck

A MAJOR PROBLEM FOR SCHOOLS

How do we incorporate what we know about the development of male and female learners to move toward more productive and inclusionary science practices and pedagogy? Today schools are striving more than ever to create educational programs that are nonsexist and multicultural in their orientation (Oakes, 1990). The concern is for building science programs that will more effectively meet the needs of all students (NSTA, 1990). The focus of this chapter is on gender issues and the effects of problem-centered teaching practices, especially STS techniques for encouraging the participation of girls and young women in science. The challenge for science educators is to examine current teaching practices in light of the research on gender and education.

THE "FIX 'EM" MENTALITY

The underachievement of girls in science has been met with cries of "What's wrong with the girls?" However, when half the population of students are turned off by science, we need to ask, "What's wrong with the science education?" Science educators have been quick to embrace stereotypic notions of the successful science student. Students who have different needs, experiences, and beliefs are often dismissed in traditional science classrooms as being not good enough (AAUW, 1992). In many situations, blame and shame are cast on the students if they do not succeed in science or choose not to participate

(Kelly, 1987). As a result, many students move away from science and achieve success in other areas.

Breaking away from this stereotypic thinking is perhaps one of the most important challenges science educators face today. It is often our beliefs and attitudes from the past that get in the way of change (Fullan, 1990; Sarason, 1990; Magolda, 1992). Many female as well as male students have been thought of as being so different from the stereotypical science student in the traditional science classroom that their potentials in science have been overlooked. The goal for many science teachers has become to "fix" the students who are different so that they fit the mold. As Sheila Tobias (1991) expresses it, science educators are looking in the "out groups" for "in-group" types. The belief that only certain types of students tend to achieve in science limits the potential for broadening participation in science, while reinforcing notions that science educators seek to produce duplicates of themselves.

Research is beginning to emerge that suggests that the molds be broken and cast aside; that the role of the teacher be examined closely. Differences among students in the science classroom should be celebrated and nurtured. Teachers must come to see themselves as guides on the side rather than sages on stages. The ultimate goal is to empower both the students and teachers through the development of their natural abilities.

We must help students develop all their thinking skills by providing experiences that challenge their current views and take them from the previous self into the developing self (Magolda, 1992; Belenky, 1986; Blunck, Giles, and McArthur, 1993; Kelly, 1987; Koch, 1993; Rosser, 1990; Wilbur, 1991). Traditional science teaching practices fall short in terms of meeting the needs of all students. In a 1992 report titled *How Schools Shortchange Girls*, published by the American Association for University Women, researchers point out many ways that girls are shortchanged in science and mathematics education (AAUW, 1992). The teaching practices that limit girls' participation in classroom science include: calling on boys more frequently, allowing girls to opt out of complex hands-on experiences, encouraging boys to solve problems on their own while "doing it" for the girls, using the male pronoun "he" to represent all scientists. These will be further addressed later.

Equity in science education implies fairness in the distribution of services, equal access to programs/courses, and the inclusion of nondiscriminatory teaching practices in science. Segregation in our schools on the basis of gender has become a legal as well as moral issue since the advent of Title IX in 1972. Even though efforts spearheaded through Title IX have tried to provide for equal access to science courses and extracurricular activities related to science, research shows students are still treated differently on a number of student attributes including gender (AAUW, 1991, 1992; Good and Brophy, 1987).

Most attempts to deal with gender effects have failed because they have focused on the "problem" population and do not deal with the dynamic complexities of learning that occur within the context of the classroom (Magolda, 1992). The remedies have often taken the form of pull-out curricula, or fragmented curricula, which involved add-on components that failed to blend with other dimensions of the curricula or address personal needs of the students (Wilbur, 1991). This lack of integration and coordination often portrays the experiences as corrective rather than nurturing. Again the notion of fixing the student to "fit the mold," stands out as the most common remedy. Many successful intervention programs on behalf of girls and science have not been mainstreamed into dominant curricula when funding runs out (Tobias, 1992).

A great deal of research has focused on "fixing" students. A number of studies point out how male and female learners differ. The majority of research and discussion on gender issues has been concerned with characterizing the learner and defining the problem. Researchers have examined psychological and developmental differences (Kelly, 1987). Kahle and Lakes (1983) found that a number of factors, including societal and parental pressures, affect student attitude toward science. It has been known for some time that female students have been shown to exhibit less positive attitudes toward science than their male counterparts (Skolnick, Langbort, and Day, 1982). Males tend to be more confident in the area of science (Kelly, 1987), and females do not perform as well as males when it comes to science (Kahle, 1983). Females between the ages of nine and fourteen lose interest in science (Hardin and Dede, 1978; National Assessment of Education Progress, 1978, 1988). What does this type of research really tell us? Certainly, this type of research fails to examine the complexities of learning that are both individual and interactive. It helps us identify differences but leaves us wondering how best to deal with these differences in the science classroom.

Celebrating Diversity—Creating a New Paradigm

Research is beginning to emerge that holds promise for a brighter future for both male and female learners with respect to science education. New models for science teaching and new curriculum design are being implemented that are based on the needs, experiences, and beliefs of the learner. The majority of these problem-solving approaches embrace the tenets of Constructivism. Central to the constructivist approach is the idea that knowledge is not passively received but actively constructed by the learner. Cognition is viewed as being adaptive; allowing for personalized organization of the material world (von Glasersfeld, 1988; Yager, 1991).

Individual student differences become the cornerstone for instruction. Students are invited to engage in active collaborative interactions with peers and their teachers nurturing a mutual responsibility for learning. Our goals for the nineties and beyond should not be centered on replacing a womanless curriculum with a manless curriculum, but rather to transform the curriculum to include everyone (NSTA, 1990). Connecting students to their science experiences should be the goal. Students should be given the opportunity to question the relationships between science and technology in a social context allowing students to assess the benefits for the environment and other human beings critically. Rosser states that adopting this perspective may be the most important change that can be made for all people, both male and female (Rosser, 1990).

Freedom of expression is encouraged and contradictory points of view are valued. This type of approach requires that teachers see themselves as "constructed knowers" and increase in their abilities to be "fluid and flexible" in teaching practice rather than relying on standard teaching formulae (Magolda, 1992; Belenky et al., 1986).

Marcia Magolda (1992) reports on an extensive, qualitative, longitudinal study of students' learning at the college level. This research is different from other pioneering work in that both males and females are part of the sample. The study traces the cognitive growth of the students through an extensive interview and coding system. Conclusions from this study suggest that learning patterns are related to but not dictated by gender. Some patterns may be used more frequently by one gender, but both genders combine approaches at different stages of their development. Patterns for both genders are equally complex and must be equally valued to create the climate where lasting learning can occur. Magolda's work moves us closer to reducing the stereotypic notions about the ways women and men learn.

Early childhood experiences traditionally provide boys with greater opportunities to build and construct models. Koch's erector set theory (1993) maintains that girls are at a disadvantage in science because they have not had equal opportunities to build models with blocks and erector sets and knock them down and build them up again. This form of early risk-taking behavior allows boys to be more comfortable with possibilities of failure in science activities, understanding that they can try again, if at first their model, activity, investigation, experiment does not work.

Other research on curriculum transformation provides a vision for bringing male and female students together as a community of learners. Emily Style (1988) uses the metaphor of windows and mirrors to support the belief that curriculum needs to provide mirrors for students in order for them to see themselves reflected in the course of study as well as providing windows into new knowledge. In most traditional science classrooms, girls do not see their experiences or the experience of women in science mirrored to them. Establishing these personal connections is an essential element in creating quality science experiences. The idea of connections is perhaps one of the most important characteristics of an inclusionary approach (Belenky et al., 1986; Koch, 1992; Style, 1988; Wilbur, 1991). In a recent study of science-avoidant college women, there was an overwhelming belief that "science did not have anything to do with the real world or with my life." This was considered the major influence in turning college women away from science and science-related fields (Koch, 1993). Connections between science and human beings are a very important concern. These connections could serve as the link to attract more women, people of color, and males not now attracted to science as it is taught in the traditional manner (Rosser, 1990). These connections provide students with the opportunity to see themselves reflected in the day-to-day experiences in the classroom. The "connected curriculum" should serve to connect students to:

1. themselves—by providing an environment that builds positive feelings toward science-related personal attributes;
2. science—by encouraging the development of a personal interests stemming from *student* questions and experiences;
3. each other—by helping students establish relationships based on the appreciation of people's talents and strengths;
4. the teacher—by creating a relationship built on personal support and mutual respect; and
5. the real world—by encouraging students to question their surroundings and get involved outside the classroom.

The goal in developing inclusionary practices should be to normalize the effect on the differences between male and female learners. Many gender issues that arise in traditional science classrooms result from exclusionary pedagogical techniques (Belenky et al., 1986; Rosser, 1990; Wilbur, 1991). Wilbur identifies six attributes of a "gender fair" approach to science teaching. A gender-fair approach:

1. should acknowledge and affirm variation;
2. should be inclusive, viewing differences within and among groups of people in a positive light. Students should see themselves reflected in the approach and identify positively with the personal messages they uncover;
3. should be accurate, helping students uncover information and ideas that are valuable and capable of withstanding critical analysis;
4. should be affirmative, emphasizing the value of individuals and groups;
5. should be representative, allowing students to uncover multiple perspectives on all sorts of issues; and
6. should be integrated, weaving together the interests, needs, and experiences of both male and female students.

When these elements are in place, national assessments have shown that student attitudes become more positive for middle and high school students especially female students (NAEP, 1978, 1988).

STS Instruction and Female Attitudes Toward Science Class

Not until recently has evidence started to emerge on the differential effects of inclusionary approaches. How are students affected by these practices? The Iowa Chautauqua Program, an inservice program for science teachers of kindergarten through twelfth grade, has been looking at just this question. The Chautauqua teachers use the STS approach in their classrooms. The Iowa definition of STS has considered and incorporated the inclusive characteristics presented in this discussion.

Before STS instruction, females exhibited more negative attitudes toward science. Some interesting changes in student attitude have been discovered. These attitude shifts reflect more positive perceptions about science in general and specific teacher characteristics. But most important, the gap between female and male learners has narrowed.

Blunck and Ajam (1991) looked at gender-related differences in students' attitudes toward science, science classes, and science teachers. The experimental design involved using pretest-posttest measure of treatment and control groups. Using data collected by twenty Iowa Chautauqua teachers, the researchers found that female students enjoyed their science classes more than males when STS practices were used. After their STS experiences, the attitudes of females shifted significantly. Female students also exhibited more positive attitudes toward their science teacher.

Perhaps the most exciting finding from this study is that STS instruction seems to be narrowing the gap that usually exists between female and male learners. Figures 11.1 and 11.2 reveal changes in student attitude related to students' perceptions of science and their science teacher.

Mackinnu (1991) has investigated the differential gender effects of STS instruction compared to a textbook approach. Over 700 students and teachers were involved in this study. Comparison of the t-tests on pretest and posttest scores showed a decrease in the number of classes with significant differences between male and female learners. "This means that STS instruction does minimize the gap between the female and male attitudes toward science for the teachers involved in this study" (Mackinnu, 1991, p. 118).

These studies represent a beginning in terms of research on inclusive constructs in science teaching. Much more research on gender effects is needed. It is imperative that science educators be collecting a wide variety of evidence

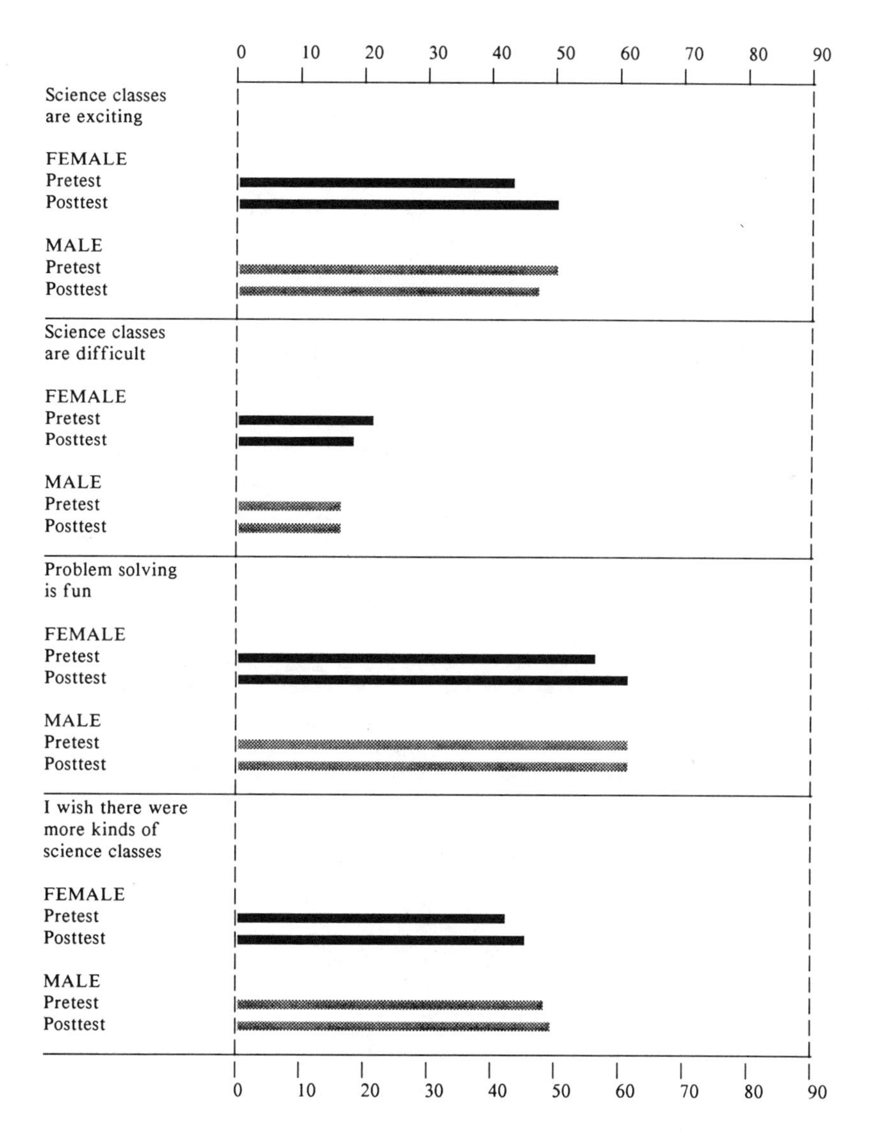

FIGURE 11.1 Students' Views of Science: Areas Where STS has Shown a Normalizing Effect Favoring Females

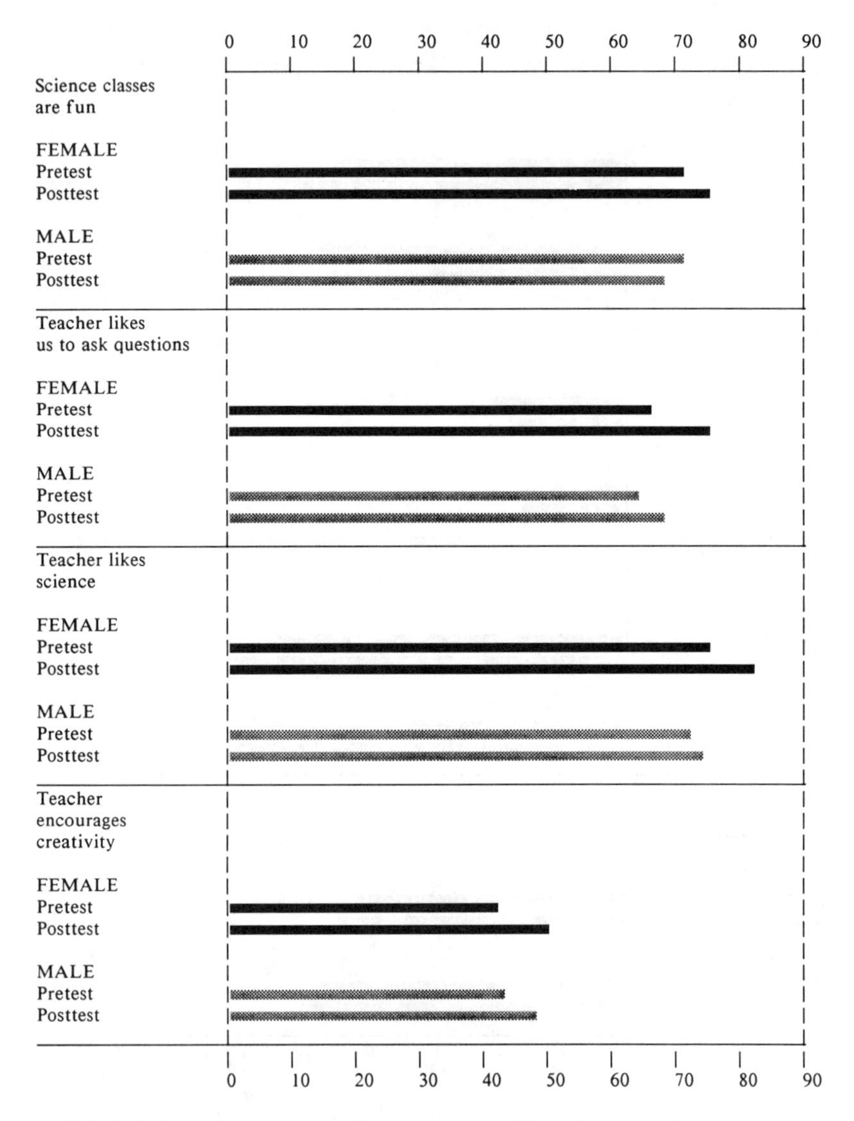

FIGURE 11.2 Students' Views of Science: Areas Where STS has Shown a Differential Effect Favoring Females

that will show the ways in which female and male learners are changing as a result of their involvement in an inclusionary learning environment. How can these environments be best described and established?

Creating Girl-Friendly Science Classes

Strategies of encouragement and inclusion can be developed by teachers at all levels with STS approaches. Koch (1992) identifies specific techniques for creating "girl-friendly" science classrooms:

1. Demonstrating the connections between science and technology and their lived experience is extremely enhancing for girls. These connections are part of the fabric of STS education and offer teachers a wonderful possibility for changing the way girls have traditionally viewed science.
2. Cooperative learning techniques have been very successful in encouraging the participation of girls in science and technology (Mastny, 1992). When cooperative learning groups are structured and planned to include a heterogeneous group with respect to race, gender, ethnicity, and ability, all students more readily contribute their talents to the problem with which the class is engaged.
3. Teachers need to listen to the small, soft voices of girls, in the face of frequently more vocal, aggressive, male voices, especially in the upper grades. Students who do not feel entitled to doing and knowing science and technology frequently pull inward and do not express themselves in class. All students need to be actively engaged in classroom conversations and all students need to hear scientists referred to from both genders. At every school level, cultural norms inhibit the identification of girls with science.
4. Frequently science teachers *do* the lab or project for the female students who ask for assistance, while encouraging the boys to figure it out on their own. Far from helping these girls, the message that is communicated is that "you are not able."
5. Science is seen as a male province where assumed female squeamishness and lack of mechanical aptitude has no place. These assumptions are an example of cultural stereotyping that prevents girls from fulfilling their potential in science and technology. Despite these girlhood stereotypes, which do not allow for girls to "get messy," it is most often the women they become who do the real "messing about" as they maintain the fabric of daily life. Teachers would be well advised to bring this gender agenda to the formal curriculum and enable girls to relate to the ways in which women traditionally "get messy."
6. People who have traditionally felt excluded from science and technology have not been socialized to believe that they can succeed in these fields.

They need to meet people in science and technology who can serve as role models and mentors. Local professional organizations are willing to provide visitors for schools and classrooms. Science educators need to make their classrooms more inviting and inclusive to compensate for the lack of role models and cultural encouragement.

Teachers often unwittingly create classroom inequities. Fortunately, gender equity workshops for science educators have enabled teachers to identify their own patterns of bias in the classroom and correct them. Teachers are often surprised that their own behaviors have not been in the best interests of all their students. Once they have learned to code their interactions with students in terms of equitable treatment and expectations, they are able to change their teaching style significantly (Griffin, 1991).

Conclusions

Attempts to reform school science programs are currently focused on making science more meaningful for all (NSTA, 1990). It is extremely important that science teachers develop a sensitivity and understanding of the needs of both the female and male learner. The development of inclusionary practices must address gender issues from a new perspective. The challenge comes in trying to move away from creating interventions to "fix" science students to accommodate traditional science programs. The focus rather should be on fixing the science programs. Students must be provided the opportunity to make a genuine contribution in their science experiences—questioning, testing, and analyzing the natural world in the context of human experience. The National Coalition of Girls' Schools Symposium on Math and Science for Girls in 1991 asserted that nurturing a sense of connectedness with the natural world and valuing the ability to question is necessary for equity to exist in science and mathematics education. Because we are all part of what we are trying to change, there needs to be a conscious and deliberate effort to transform traditional science curriculum and change teaching styles and classroom culture to include us all.

References

American Association for University Women. (1991). *Shortchanging girls: Shortchanging America.* Washington, D.C.: American Association of University Women.

American Association for University Women. (1992). *How schools shortchange girls.* Washington, D.C.: American Association of University Women.

Belenky, M. F., Clinchy, B. M., Goldberger, N. R., and Tarule, J. M. (1986). *Women's ways of knowing*. New York: Basic Books.

Blunck, S. M., and Ajam, M. (1991). Gender-related differences in students' attitude with STS instruction. *Chautauqua Notes, 6*(2), 2–3.

Blunck, S. M., Giles, C. S., and McArthur, J. M. (1993). Gender differences in the science classroom: STS bridging the gap. In R. E. Yager (ed.), *What research says to the science teacher, Volume 7: The science, technology, society movement* (pp. 153–160). Washington, D.C.: National Science Teachers Association.

Good, T. L., and Brophy, J. E. (1987). *Looking in classrooms*. New York: Harper and Row.

Griffin, K. R. (1991). Attending to equity. *AAUW Outlook, 23*(4), 10–15.

Hardin, J., and Dede, C. J. (1978). Discrimination against women in science education. *The Science Teacher, 40*, 18–21.

Kahle, J. (1983). *The disadvantaged majority: Science education for women*. Burlington, N.C.: Carolina Biological Supply Company. AETS Outstanding Paper for 1983.

Kahle, J. B., and Lakes, M. K. (1983). The myth of equality in science classrooms. *Journal of Research in Science Teaching, 20*, 131–140.

Keller, E. F. (1982). Feminism and science. *Signs, 7*(3), 589–602.

Kelly, A. (1987). *Science for girls?* Philadelphia: Open University Press.

Koch, J. (1992). Tips for teachers: Science is for everyone. In A. Mastny (ed.), *Science Teams*. New Brunswick, N.J.: Rutgers Consortium for Educational Equity.

Koch, J. (1993). *Lab coats and little girls: The science experiences of women majoring in biology and education at a private university*. Ann Arbor: University Microfilms International, no. 9317670.

Mackinnu. (1991). *Comparison of learning outcomes between classes taught with a science-technology-society (STS) approach and a textbook oriented approach*. Unpublished doctoral dissertation, University of Iowa, Iowa City.

Magolda, M. B. (1992). *Knowing and reasoning in college: Gender-related patterns in student's intellectual development*. San Francisco: Jossey-Bass.

Mastny, A. (1992). Cooperative learning teaching techniques. In A. Mastny (ed.), *Science Teams*. New Brunswick, N.J.: Rutgers Consortium for Educational Equity.

National Assessment of Educational Progress. (1978). *The third assessment of science, 1976–77*. Denver: Author.

National Assessment of Educational Progress. (1988). *The science report card: Elements of risk and recovery*. Princeton: Educational Testing Service.

National Coalition of Girls' Schools. (1992). *Math and science for girls*. Concord, Mass.: National Coalition of Girls' Schools.

National Science Teachers Association. (1990). Science/technology/society: A new effort for providing appropriate science for all. (The NSTA position statement.) *Bulletin of Science, Technology & Society, 10*(5 & 6), 249–250.

Oakes, J. (1990). *Multiplying inequalities: The effects of race, social class and tracking on opportunities to learn mathematics and science*. Santa Monica, Calif.: The RAND Corporation.

Rosser, S. V. (1990). *Female-friendly science: Applying women's studies methods and theories to attract students*. New York: Pergamon Press.

Skolnick, J., Langbort, C., and Day, L. (1982). *How to encourage girls in math and science: Strategies for parents and educators*. Englewood Cliffs, N.J.: Prentice-Hall.

Style, E. (1988). Curriculum as window and mirror. In *Listening for all voices: Gender balancing the school curriculum*. Summit, N.J.: Oak Knoll School.

Tobias, S. (1991). *They're not dumb, they're different: Stalking the second tier*. Tucson: Research Corporation of America.

Tobias, S. (1992). *Revitalizing undergraduate science: Why some things work and most don't*. Tucson: Research Corporation of America.

von Glasersfeld, E. (1988). *Cognition, construction of knowledge, and teaching*. Washington, D.C.: National Science Foundation.

Wilbur, G. (1991, August). *Gender-fair curriculum*. Research report prepared by Wellesley College Research on Women.

Yager, R. E. (1991). The constructivist learning model: Towards real reform in science education. *The Science Teacher, 58*(6), 52–57.

CHAPTER 12

ADVANTAGES OF STS FOR MINORITY STUDENTS

Joan Braunagel McShane
Robert E. Yager

DIFFERENCES BETWEEN MINORITY AND MAJORITY STUDENTS

In the United States, blacks, Hispanics, and Native Americans make up approximately 18 percent of the population, but comprise only 2.2 percent of the science and engineering work force (Malcolm, 1985). According to NSF (1984) estimates, females earn 12.2 percent of the doctoral degrees, 24.8 percent of the master's degrees, and 15.3 percent of the bachelor's degrees in science when employment in the labor force was studied. An NSF report (1980) suggests that males without graduate degrees find careers in science more easily than females with either a master's or a doctorate.

Racial minorities and females have been consistently underrepresented in mathematics and science majors and careers for at least the last five decades (Hill, Pettus, and Hedin, 1990). This fact has led to a growing national concern for increasing the participation of minorities and females in science and technology careers (NSF, 1984).

Researchers concerned with cultural diversity and multicultural education maintain that certain people of color differ in their worldviews and cognitive styles from those held by the dominant culture (Anderson, 1988; Banks and Banks, 1989; Banks and Lynch, 1986; Burgess, 1986; Gollnick and Chinn, 1990; Hale, 1986; Kagan and Madsen, 1971; Ramirez, 1978; Sleeter and Grant, 1988). These researchers compare the philosophical worldview of certain minority groups with their nonminority counterparts in a variety of dimensions. Table 12.1 provides a summary of this research.

TABLE 12.1 Worldview Comparisons of Persons from Western and Non-Western Cultures

Non-Western	*Western*
1. Emphasis on group cooperation	1. Emphasis on individual competition
2. Achievement as it reflects group	2. Achievement for the individual
3. Holistic thinking	3. Dualistic thinking
4. Socially oriented	4. Task oriented
5. Extended Family	5. Nuclear family
6. Accept affective expression	6. Limited affective expression
7. Religion permeates culture	7. Religion distinct from other parts of culture
8. Time is relative	8. Rigid time schedule
9. Value harmony with nature	9. Mastery and control over nature

FEATURES OF STS CLASSROOMS

Within the context of STS classrooms, the cognitive structure of minority students is accommodated through cooperative learning and assessed through the use of portfolios. The integrative, topical curriculum and in-depth study of subject matter characterized by NSTA, science educators, and researchers (Rubba, 1989; Rubba, McGuyer, and Wahluen, 1991; Spector and Gibson, 1991; Wraga and Hlebowitsh, 1991; Yager 1991) is consistent with several of the philosophical worldviews of minority students mentioned above, particularly with respect to holistic thinking and integrative disciplines. Also, the STS rationale parallels cognitive psychologists' theories regarding the role of prior knowledge in cognitive structure (Ausubel, 1968; Champagne and Klopfer, 1984; Resnick, 1986) and the effect that sequential organization of subject matter has on the stability and clarity of anchoring ideas for subsequent learning (Ausubel, 1968). The incorporation of new concepts and information into an existing and established cognitive framework is largely influenced by the student's past experiences and the integrative nature of subject-matter discipline (Ausubel, 1968; Champagne and Klopfer, 1984; Yager and McCormack, 1989).

The research associated with minorities adds an interesting dimension to the needed reforms in science education. Certainly the findings provide information about necessary changes. However, when the goals have clearly shifted to serving *all* students better, special cases and programs may not be as vital for females and minority students. There is considerable evidence that current programs are serving no one well, including students who are most successful and those who pursue further study of science in college.

Since the STS approach has been found to serve most students better than traditional approaches, a look at its specific effect on minority students was undertaken. Although there are not large minority enrollments in Iowa schools generally, Davenport is a district where significant numbers are found. Also, two Lead Teachers associated with the Iowa Chautauqua Program use the STS approach in upper-elementary grades and the middle school.

Joan McShane teaches science in grades four through six at Jefferson School where minority enrollment is about 55 percent. Chris Hull is a ninth-grade teacher at Central High School where low-ability students are involved with STS. Some sections include many problem youth where teacher challenges are constant. Chris's successes with STS instruction provide some insights concerning its effectiveness.

There are eight features that characterize ideal STS teaching. McShane and Hull were observed to excel in all categories. These include the features indicated in Table 12.2.

Differences between Minority and Majority Students when Experiencing STS Instruction

Results with STS instruction indicate the following changes in all students, including minorities and females (Yager, 1993; Yager and Penick, 1990):

1. increased attention span;
2. a gradual shift from concrete to abstract functioning;

TABLE 12.2 A Contrast of STS Classrooms and Traditional Science Classrooms

STS	*Traditional*
1. Class activities are student-centered.	1. Class activities are set and controlled by the teacher.
2. Individualized and personalized, recognizing student diversity.	2. Group instruction geared for the average student.
3. Directed by student questions and experiences.	3. Directed by the textbook.
4. Uses a variety of resources.	4. Uses basic textbook almost exclusively.
5. Cooperative work on problems and issues.	5. Some group work, primarily in laboratory.
6. Students are considered active contributors to instruction.	6. Students are seen as recipients of instruction.
7. Teachers build on student experiences, assuming that students learn best from their own experiences.	7. Teachers do not build on students' experiences, assuming that students learn more effectively by being presented with organized, easy-to-grasp information.
8. Teachers plan their teaching around problems and current issues.	8. Teachers plan their teaching from the prescribed curriculum guide and textbook.

3. greater ability to comprehend and manipulate abstractions and to see relationships between abstractions;
4. higher levels of abstractions, generality, and inclusiveness;
5. the ability to discriminate between two sets of analogous ideas; and
6. the ability to transfer bodies of knowledge.

STS approaches enable minority students to achieve as well as majority students. The results of a comparison of successes with minority and majority students are displayed in Figures 12.1 and 12.2. It is apparent that STS is effective in reducing differences in learning in the five domains tested for minority students.

STS instruction seems to bridge gaps where minority effects are a concern. The differences between the successes of majority and minority students can be seen by comparing the data in Figure 12.2. As we move toward more inclusionary practices in science teaching, STS stands out as a powerful alternative for making science more meaningful for all students.

Some Generalizations

STS has been shown to be superior to traditional teaching in a variety of studies and in various goal and assessment domains (Iskandar, 1992; Mackinnu,

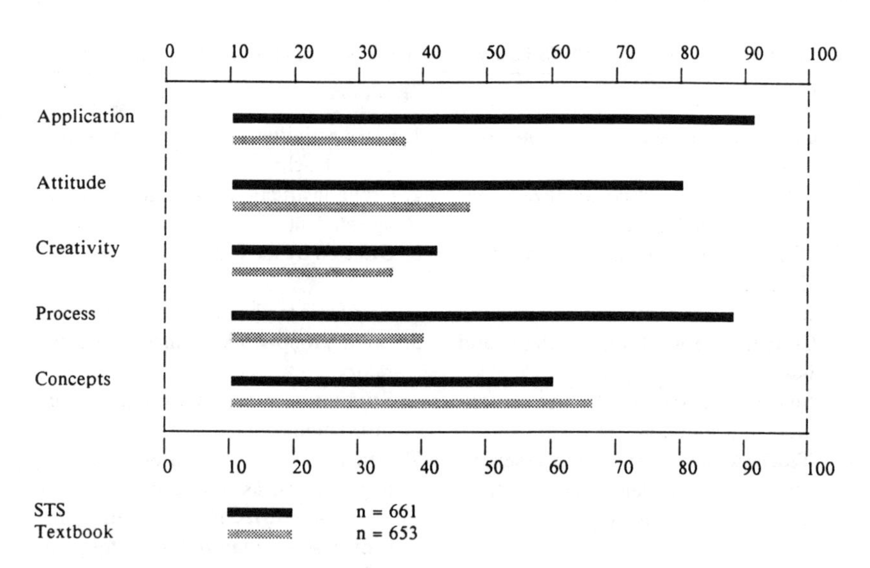

FIGURE 12.1 Comparison of Student Growth in the Five Assessment Areas for Those Enrolled in Textbook and STS Sections Taught by Fifteen Lead Teachers in the Iowa Chautauqua Program.

1991; McComas, 1989; Myers, 1988). In fact, the following generalizations are possible when considering the advantages of STS instruction for minority students:

1. The results when differences are noted between minority and majority students in STS sections produced major differences. Minority students per-

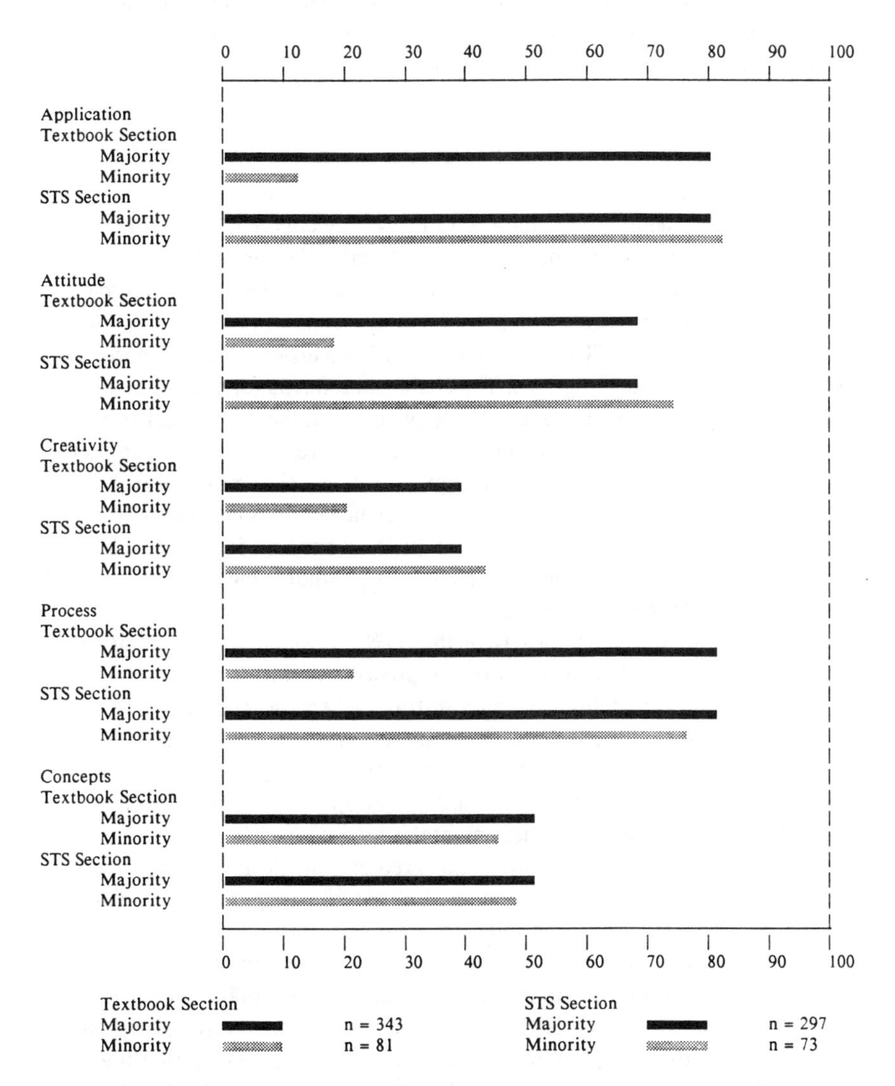

FIGURE 12.2 Comparison of Minority and Majority Student Growth on Scales in Five Assessment Areas for Students Enrolled in Textbook and STS Sections Taught by Two Lead Teachers in Davenport.

form better than majority students in the application, attitude, and creativity domains, while majority students score better in the process and concept domains which are the foci in traditional classrooms. These differences are striking since majority students are favored in typical science classrooms.

2. The results in non-STS (textbook) sections where minority and majority students are studied illustrate the advantages of textbook teaching for majority students, which is reversed in STS sections.

SUMMARY

Given the fact that the majority of researchers agree that school science programs must be transformed and restructured to better meet the needs of minority students, STS is emerging as a viable alternative to traditional science programs. Given the research that is appearing recently in the professional literature, the hope is that STS will continue to make a difference for minority students in school science programs. The challenge remains for science educators who are using STS to continue to collect evidence on the differential effects of their instruction with minority students. Too often we are quick to dismiss the idea that minority issues still exist within our science classrooms. It is the sensitivity of the teacher to these issues in the science classrooms that will, in the long run, make the biggest difference of all. As we search for more inclusionary approaches to science teaching, STS must be studied carefully in terms of its positive effects on all learners.

Studies of the successes with the STS approach with minority students must be extended. Several of the positive results with minority students cry for corroboration in urban settings where instructional problems are more intense. Nonetheless, the research is clear; when students help define the issues, locate needed resources, debate the validity of information and expert judgments, arrive at class (or cooperative group) consensus, and take corrective actions, their mental engagement is such that learning follows automatically. Of course, the greatest hurdle is getting involvement in terms of student questions and individual engagement with finding answers. When the human mind is challenged and engaged, learning is inevitable. And yet, the human mind is rarely engaged when teachers determine the curriculum, the study units, the daily lesson plans. Teachers can teach but students learn little unless they are directly involved. STS demands such involvement and that may be the major reason that the STS approach is successful with minority students where more traditional instructional approaches fail.

References

Anderson, J. A. (1988). Cognitive styles and multicultural populations. *Journal of Teacher Education, 39*, 2–9.

Ausubel, D. P. (1968). *Educational psychology: A cognitive view.* New York: Holt, Rinehart & Winston.

Banks, J. A., and Banks, C. (1989). *Multicultural educational issues and perspectives.* Boston: Allyn and Bacon.

Banks, J. A., and Lynch, J. (1986). *Multicultural educational in western societies.* New York: Praeger.

Burgess, B. J. (1986). Native American learning styles. In L. A. Morris, G. Sather, and S. Scull (eds.), *Extracting learning styles from social/cultural diversity: Studies of five American minorities.* Norman, Okla.: University of Oklahoma Press.

Champagne, A. B., and Klopfer, L. E. (1984). Research in science education: The cognitive psychology perspective. In D. Holdzkom and P. B. Lutz (eds.), *Research within reach: Science education* (pp. 171–189). Charleston, W.V.: Research and Development Interpretation Service, Appalachia Educational Laboratory.

Gollnick, D., and Chinn, P. (1990). *Multicultural education in a pluralistic society* (3d ed.). New York: Merrill.

Hale, J. (1986). Cultural influences on learning styles of Afro-American children. In L. A. Morris, G. Sather, and S. Scull (eds.), *Extracting learning styles from social/cultural diversity: Studies of five American minorities.* Norman, Okla.: University of Oklahoma Press.

Hill, O. W., Pettus, W. C., and Hedin, B. A. (1990). Three studies of factors affecting the attitudes of Blacks and females toward the pursuit of science and science related careers. *Journal of Research in Science Teaching, 27*(4), 289–314.

Iskandar, S. M. (1991). *An evaluation of the science-technology-society approach to science teaching.* Unpublished doctoral dissertation, University of Iowa, Iowa City.

Kagan, S., and Madsen, M. C. (1971). Cooperation and competition of Mexican-American and Anglo-American children of two ages under four instruction sets. *Developmental Psychology, 5*, 32–39.

Mackinnu. (1991). *Comparison of learning outcomes between classes taught with a science-technology-society (STS) approach and a textbook oriented approach.* Unpublished doctoral dissertation, University of Iowa, Iowa City.

Malcolm, S. (1985). The place of minorities in the science and technical work force. *Black Scholar, 16*, 50–55.

McComas, W. F. (1989). Just the facts: The results of the 1987–88 Chautauqua workshops. *Chautauqua Notes*, *4*(4), 1–2.

Myers, L. H. (1988). *Analysis of student outcomes in ninth grade physical science taught with a science/technology/society focus versus one taught with a textbook orientation.* Unpublished doctoral dissertation, University of Iowa, Iowa City.

National Science Foundation. (1980). *U.S. scientists and engineers, 1978.* (NSF 80-304). Washington, DC: U.S. Government Printing Office.

National Science Foundation. (1984). *Women and minorities in science and engineering.* Washington, D.C.: U.S. Government Printing Office.

Ramirez, M. (1978). Cognitive styles and cultural democracy in education. *Social Science Quarterly*, *53*, 895–904.

Resnick, L. B. (1986). *Cognition and instruction: Theories of human competence and how it is acquired.* Pittsburgh: Learning Research and Development Center.

Rubba, P. A. (1989). An investigation of the semantic meaning assigned to concepts affiliated with STS education and of STS instructional practices among a sample of exemplary science teachers. *Journal of Research in Science Teaching*, *26*(8), 687–702.

Rubba, P. A., McGuyer, M., and Wahluen, T. M. (1991). The effects of infusing STS vignettes into the genetics unit of biology on learner outcomes in STS and genetics: A report of two investigations. *Journal of Research in Science Teaching*, *28*(6), 537–552.

Sleeter, C., and Grant, G. (1988). *Making choices for multicultural education.* New York: Merrill.

Spector, B. S., and Gibson, C. W. (1991). A qualitative study of middle school students' perceptions of factors facilitating the learning of science: Grounded theory and existing theory. *Journal of Research in Science Teaching*, *28*(6), 537–552.

Wraga, W. G., and Hlebowitsh, P. S. (1991). STS education and the curriculum field. *School Science and Mathematics*, *91*(2), 54–59.

Yager, R. E. (1991). The constructivist learning model: Towards real reform in science education. *The Science Teacher*, *58*(6), 52–57.

Yager, R. E. (1993). Science/Technology/Society and teacher education: Solutions for problems of diversity and equity. In M. J. O'Hair and S. J. Odell (eds.), *Diversity in teaching: Teacher education, Yearbook 1* (pp. 243–268). New York: Harcourt, Brace, Jovanovich.

Yager, R. E., and McCormack, A. J. (1989). Assessing teaching/learning successes in multiple domains of science and science education. *Science Education*, *73*(1), 45–58.

Yager, R. E., and Penick, J. E. (1990). Science teacher education. In W. R. Houston (ed.), *Handbook of research on teacher education* (pp. 657–673). New York: Macmillan.

Chapter 13

Enhancing Student and Teacher Understanding of the Nature of Science Via STS

Lawrence R. Kellerman
Chin-Tang Liu

The views of science as historically rooted and as an enterprise affected by society and culture have not influenced science education. This is the case despite the serious reform efforts of the 1960s and 1970s to develop new curriculum materials and to provide corresponding teacher education. Nor were the reform efforts of the 1980s any more successful, even with new research efforts to determine instructional effects on student learning. However, the STS reform movement may have the potential for influencing science education since it focuses on changing the goals and content of science teaching in ways that will link science to social concerns.

Research concerning "teaching and learning about the history and nature of science" has gained much attention in science education in recent years as evidenced by two International Symposia: History, Philosophy, and Science Teaching held at Florida State University, and Queen's University, Kingston, Ontario, Canada, in 1989 and 1992 respectively. Moreover, the advocacy of the development in scientific literacy from National Science Teachers Association (NSTA, 1982), American Association for the Advancement of Science (AAAS, 1989), and the National Assessment of Educational Progress (National Center for Education Statistics, 1993) emphasizes the need for a more accurate understanding of science and technology as a critical component for reform.

Alarming calls based on review of the previous research work generally indicate "students do not possess adequate knowledge about the nature of science" (Lederman, 1992, p. 335). Whether preservice and inservice teachers have an appropriate and accurate understanding of the nature of science continues as a debate.

A broader view of the nature of science has been used in Iowa STS research involving teachers and their students. The nature of scientific knowledge, the nature of the scientific enterprise, and the nature of scientists has been discussed—mostly as goals of typical science instruction not realized with current materials and approaches (Cooley and Klopfer, 1963; Kimball, 1968). Project 2061 (Rutherford and Ahlgren, 1989) identifies the nature of science as one of four categories of knowledge, skills, and attitudes essential for all citizens in a scientifically literate society. Three major components of the nature of science, as defined by the AAAS (1989), are:

1. scientific world view—the world is understandable, scientific ideas are subject to change, scientific knowledge is durable, and science cannot provide complete answers to all questions;
2. scientific methods of inquiry—science demands evidence, science is a blend of logic and imagination, science explains and predicts, scientists try to identify and avoid bias, and science is not authoritarian; and
3. nature of the scientific enterprise—science is a complex social activity, science is organized into content disciplines and is conducted in various institutions, there are generally accepted principles in the conduct of science, and scientists participate in public affairs both as specialists and as citizens.

In this chapter, we report on STS stories that were produced in the Iowa Chautauqua Program and Iowa Scope, Sequence, and Coordination Project that illustrate the current efforts to succeed in the nature of science domain in STS classrooms. In Iowa, the STS approach is advocated in science classrooms where the expectation is that teachers and students will improve in terms of their understanding of the nature of science.

Stories of Science in the STS Classroom

STS means focusing on problems, on questions, on unknowns. It is a never-ending process. More importantly, it seems on the surface to reflect how scientists do science, and it does it within the classroom setting. Examples gathered over the previous decade show that STS instruction, when practiced by students in the classroom, reflects real science in the laboratory. Kellerman (1993*a*) has provided a portrait of one teacher's attempt to bring STS instruction into his classroom. This effort resulted in students spending an entire year studying the issue of ozone depletion, gaining or reinforcing their ability to perform research, critical thinking, and problem solving.

More importantly, these students were able to "do science." The students wrote to companies manufacturing foam products and became familiar with the manufacturing process and the chemical reactions. One group of students examining water quality within this unit went downstream from a packing plant at 1 A.M. to gather water samples. When asked why they went out at such a time they responded, "We didn't want them to see us because we want to ask them some questions and see if they're telling the truth" (Kellerman, 1993*b*, p. 142).

Both of the aforementioned actions reflect authentic science. In both cases students responded to self-generated questions, examined the phenomenon without a formal hypothesis, and performed the activities as "science in the making." The results of the ozone depletion study were equally "science." Students shared their newfound understanding of the ozone with concerned citizen groups and provided presentations to other students. The results were not seen as "the Truth"; rather they were seen as increasing the awareness of those who heard the presentations. Kellie Tech, one of the students, noted "Maybe people will think next time when they buy some of these products" (Kellerman, 1993*b*, p. 144). What an admirable goal for a young and budding scientist!

Other examples abound that portray students "doing science" as if they were scientists. Presented here is another portrait observed while performing the formative evaluation for the Scope, Sequence, & Coordination Project-Iowa (Kellerman, 1993*a*). The observation took place in an eighth-grade classroom and is a representative portrayal of many such observations made during the course of three years of work.

An experienced Mason City (Iowa) instructor, in his thirty-seventh and last year of teaching, had been a part of the SS&C project since its inception and was one who had been trained in the STS methodology (the Chautauqua Program). This instructor and the other eighth-grade science teachers at John Adams Middle School had started an SS&C module *Nightmare on Elm Street* the day of the observation. The module was designed to get the students involved in building a structure that could one day be their home. The study included physics, biology, and earth science. The previous day he had piqued the students' interest in the topic by displaying a model house provided by the high school vocational technology class and asking the students to design their dream house as a homework assignment. The first fifteen minutes of the observation the students discussed with the other members of their group the establishment of the main points of their drawing and polishing any possible weak points. The instructor had devised a pretest that he then administered to the students (it was a combination essay/hands-on assessment and could be accomplished with or without the calculator). He also wanted the students to complete the assessment individually, even though they would be working in groups of four on their domicile. Following the assessment (he gave the students about 20 minutes), he then picked up their wish list for potential group members (each student could pick one; he would pick the other two) and discussed a couple of the questions on the assessment instrument. The author walked around the classroom prior to the assessment and interacted with the students and found most of them to be very enthusiastic. Initial comments noted by one author were:

> The dream houses were very imaginative, although few had thought of the direction the house would face, the need for water, the various requirements on the structure, and similar details. Most had thought of where they would build it, either in Beverly Hills, Wyoming, New Mexico, or other sites. (One girl informed me she was going to be a lawyer after attending Harvard and would have her house in Hawaii!)

The teacher noticed during his movement around the room that the students were having a real struggle with determining the dimensions they thought were realistic for their dream house. He decided to ask the class what they thought the dimensions of the classroom were. Over two-thirds of the students quickly raised their hands, but only one actually did anything to solve the problem. One young man immediately began to look at the cement blocks that made up the walls. He counted the ones along the north wall and the ones along the east wall and, after what seemed to be a brief mental calculation, raised his hand. The teacher, in true constructivist fashion, had waited this entire time, observing each student's reaction. He initially called on Derrick, who then gave his response to the problem. The instructor called on several more students for their responses; each of these additional responses were obvious guesses. He then told the class the approximate dimensions and Derrick had only been off by one foot in either direction. It was a good view of a constructivist teacher and a rewarding portrait of an STS classroom (Kellerman, 1993*a*, pp. 13–14).

In this particular case both the students and the instructor were reaping the benefits of STS instruction. The instructor benefited because the students enjoyed what they were doing and they provided evidence of their immature knowledge of home construction by their responses to his previous assignment and the preassessment. The students benefited because they attempted to do science within the context of a meaningful investigation. Both were developing a perception of science as an endeavor carried out by human beings and that they, as scientists are a part of the world, and not apart from it (Cleminson, 1990, p. 438). In short, STS appears once again to reflect some of the aspects of "real science."

Is the following an example of a classroom where the students' experience reflected the nature of science? In the early spring of 1986 one seventh-grade class of the three being taught by a teacher (following two five-week summer workshops providing training in STS instruction) had decided to study the extinction of dinosaurs. The issue came to the fore following a number of articles and television news reports suggesting the dinosaurs had become extinct because of the impact of a large meteor or comet on the Earth's surface and the resulting ejected material. The class, after a great deal of research and debate, decided this was most likely the best explanation for the destruction of an entire form of animal life. They based their decision on the students' presentation of evidence and the resulting debate.

However, there was one group of three boys who refused to believe that this could be the lone cause for the dinosaurs' disappearance. They had determined this scenario did not fit what seemed to be known about dinosaurs. What happened, they asked, to those dinosaurs that did not need the heat that was supposed to have decreased because of the increased cloud cover? Why would the dinosaurs that were meat eaters die out? If the vegetarian dinosaurs did disappear because of the lack of vegetation, why would the carnivorous animals disappear? Those animals should have had plenty of food. This group's conclusion, given the same information that the other students had had available to them, was that the dinosaurs disappeared because of a virus or bacterium that infected the herbivores and was passed to the carnivores over time. The students had all seen and read the same data, they had all carried out their debates and discussions in the same classroom, yet one group of three saw something in the data that resulted in a different construction.

This group maintained their opinion even through some rather tough grilling and some characteristic seventh-grade ribbing. The following year one of the science magazines that made its way into the classroom was the source for some initial corroborating information for that group's hypothesis. Scientists had developed an alternative scenario to the "extinction by cometary impact" theory so prevalent just a short time earlier. The alternate scenario had as its core the disappearance of dinosaurs by a bacterium that swept through the dinosaur population, starting with the herbivores and continuing through the food chain. The students who made a similar claim were made aware of this account and were quite proud of themselves. To them it legitimized the work they had completed and gave them a boost of self-confidence. This reflected the nature of science.

There are certain issues that follow from this account of "science." Why should the students' alternate conception be given any value? What evidence is there that confirms or counters this notion? Yes, a group of scientists might agree with these three boys, but does that mean they are right? If they are found to be mistaken, what will those who accepted the currently prevailing theory think about that particular episode? Indeed, if the class had as its focus the content of the science presented in the class, that episode may cement some students' belief that science is best left to the experts. But in this particular instance and in many others that have taken place since then in other classrooms, the focus should be less on that content and more on the processes behind that content. One cannot expect the students in any classroom to assume the understanding of the processes of science when teaching only the basic content included in most textbooks. Students need to "see and do" the processes by which data are gathered, organized, interpolated, and interpreted within a scientific investigation in order to develop an appropriate conception of science. They are able to do that in an STS classroom.

If one can envision social and cognitive interaction, debate, and argument, occurring in scientific investigations, is it also possible to envision those same characteristics in the science classroom? That these behaviors occurred has been established in many previous studies. Is it likely that the science curriculum will be allowed to include investigations where such behaviors occur? This is a very real issue and is sure to spark debate.

To some extent, the issue seems to be one of "control" vis à vis the teacher and the classroom. Whereas teachers may acknowledge that many of the behaviors noted as common in the examples reported and examined here because they are basically human behaviors and probably do occur within the context of idealized investigations, they are often hesitant to allow such behaviors to occur in the classroom. In the classroom discussions noted earlier about the demise of the dinosaurs, the students became quite boisterous and the arguments came down for a time to the supposed "stupidity" of the group of three boys. Yet through it all the students did not lose their attentiveness to the study. They indeed seemed glad to have the chance to act as if the debate they were participating in meant something instead of simply verifying something already known. For many teachers such activity and such attitudes cannot be tolerated; thus, the science curriculum in their classrooms will not allow the students to experience the nature of science.

The Iowa Chautauqua Program advocates assessing students' learning about the nature of science. Lead Teachers (those who attend leadership after successful trials with STS and then help with instruction with new teachers) conducted an action research project that produced interesting results that illustrate quantitatively what STS instruction can accomplish in this domain. These leaders constructed a scale with general descriptors of the nature of science. These included:

1. Science means questioning, explaining, and testing;
2. Science means studying the concepts produced and known by scientists;
3. Science means working with various objects and materials in classrooms and laboratories;
4. Science deals with activities that affect living, that is, in homes, schools, communities, and nations; and
5. Science is a human activity that involves acting on questions about the universe.

EXPERIMENT RESULTS

The results of this investigation of instructional success are reported in Tables 13.1 through 13.3. The statistical results permit one to conclude that the STS approach produces students who better understand the nature of science. Analyses of covariance were used to compute differences statistically using the pretest scores as the covariate. Since all the F-values were significant at the 0.01 level of significance, it can be stated that students (female and male) in STS classrooms improve in their understanding of the nature of science to a greater

TABLE 13.1 Differences between the STS and Non-STS Approach with Respect to Teacher Understanding of the Nature of Basic Science

		Mean		*ST.D.*			
Group	*N*	*Pre*	*Post*	*Pre*	*Post*	*F**	*P**
STS	160	26.91	41.20	2.18	3.32	62.17	0.00
Non-STS	56	26.83	27.25	2.13	2.25		

*F,P: F- and P-Values of analysis of variance with repeated measures between posttest scores in the STS Group and Non-STS group.

TABLE 13.2 Comparisons of Student Views of the Meaning of Science in STS and Non-STS Classrooms

		Pretest		*Posttest*		
Group	*N*	*Mean*	*S.D.*	*Mean*	*S.D.*	*F1**
A: Questioning, explaining, and testing						
STS	362	3.06	0.34	3.73	0.48	26.72†
Non-STS	360	3.04	0.33	3.09	0.33	
B: Studying the concepts produced and known by scientists						
STS	362	1.12	0.39	2.68	0.53	32.19
Non-STS	360	1.39	0.49	1.54	0.48	
C: Working with various objects and materials in classrooms and laboratories						
STS	362	2.10	0.61	2.32	0.70	2.36
Non-STS	360	2.09	0.52	2.12	0.64	
D: Affecting quality of living						
STS	362	2.01	0.65	3.09	0.59	24.62
Non-STS	360	1.08	0.64	3.11	0.67	
E: Acting on questions about the universe						
STS	362	2.62	0.75	3.19	0.69	13.33
Non-STS	360	2.57	0.71	2.71	0.71	

* Analyses were computed by Analyses of Covariance (ANCOVA).
† Significant at 0.05 level.

TABLE 13.3 Results of Gender Effect Performance of STS and Non-STS Students

	STS (N = 362)				*Non-STS (N = 360)*						
	Pre		*Post*		*Pre*		*Post*		*F1*	*F2*	*F3*
	Mean	*S.D.*	*Mean*	*S.D.*	*Mean*	*S.D.*	*Mean*	*S.D.*	*(P1)*	*(P2)*	*(P3)*
Male	9.7	2.1	10.9	2.0	9.8	1.8	10.1	1.9	24.8	13.1	19.5
Female	9.3	1.99	12.0	1.9	9.5	2.0	10.7	1.9	(0.00)	(0.01)	(0.01)

* Possible Maximum Score: 20; the score of negative items (Items 2 and 3) being inverted.

F1 (P1): F- and P-values of Analysis of Covariance between female student posttest scores in STS and non-STS classes with the pretest scores used as the covariate.

F2 (P2): F- and P-values of Analysis of Covariance between male student posttest scores in STS and non-STS classes with the pretest scores used as the covariate.

F3 (P3): F- and P-values of Analysis of Covariance between female and male student posttest scores in STS with the pretest scores used as the covariate.

degree than students (female and male) in non-STS classes. More important, female students in STS classes developed more significantly positive results with respect to the nature of science than did male students in STS classes. The results of this investigation indicate that students in STS classrooms developed a more appropriate view of the nature of science. This appears to substantiate the qualitative findings reported above.

CONCLUSION

Is there a reform effort that appears most likely to mirror the nature of science as described in the initial section of this chapter? It appears that the *Benchmarks for Scientific Literacy* (AAAS, 1993) provides the impetus for such a curriculum to be developed. The ideas are in place and the suggestions are timely and pertinent.

The Content Core (NSTA, 1992) document for SS&C appears less amenable to the incorporation of the nature of science in the science classroom. There is no call for students to initiate their own investigations, no call for teachers to let the students work as if they are scientists, no suggestion that students should wrestle with the data, argue over the interpretations or the conclusions, investigate further questions. The end result for students in a typical "pure" SS&C classroom may be better science learning, but it may not lead to a better understanding of the nature of science.

The link between the nature of science and science education does not have to rely on local or national reform efforts, however. The examples presented above show that individual teachers can allow "real science" to become part of the classroom. The fact that the examples reported occurred in an STS setting provides initial evidence that the characteristics of an STS classroom mirror to some extent the places where scientists work.

The key appears to be teacher understanding the nature of science and having the willingness to allow disagreement and dissention to occur. In an STS classroom it appears possible that this occurs. The group of students studying the issue of ozone depletion recognized that controversy may arise if their collection of water samples downstream from a packing plant was observed. The students deciding where to build their houses knew they had the freedom to "dream" about their future domicile and establish a reason for their decision. The group of boys that chose not to follow the majority in explaining the demise of the dinosaurs was willing to "take the heat," again a behavior that is a part of science. Once science teachers can get past the notion that dissent is bad, good science can become a part of the science classroom. Such a view of classroom science seems central to an STS classroom.

The need to fashion the science curriculum to reflect the nature of science can provide a positive response to the question posed at the beginning of the chapter. That question, "What is the science we want our students to be doing, and how can that science become a part of the science classroom?" has been addressed by the examples and the action research reported. It is now up to science educators to ensure that science classrooms of today and tomorrow reflect the nature of science. The need for scientifically literate citizens demands that it be accomplished, and soon.

REFERENCES

American Association for the Advancement of Science. (1989). *Science for all Americans: Summary—Project 2061*. Washington, D.C.: Author.

American Association for the Advancement of Science. (1993). *Benchmarks for science literacy: Part I: Achieving science literacy: Project 2061*. Washington, D.C.: Author.

Cleminson, A. (1990). Establishing an epistemological base for science teaching in the light of contemporary notions of the nature of science and of how children learn science. *Journal of Research in Science Teaching*, *27*(5), 429–445.

Cooley, W., and Klopfer, L. (1963). The evaluation of specific educational innovations. *Journal of Research in Science Teaching*, *1*, 73–80.

Kellerman, L. R. (1993*a*). *Final report formative evaluation: Scope, sequence & coordination–Iowa*. Iowa City: The University of Iowa, Science Education Center.

Kellerman, L. R. (1993*b*). An issue as an organizer: A case study. In R. E. Yager (ed.), *What research says to the science teacher, Volume 7: The science, technology, society movement* (pp. 141–146). Washington, D.C.: National Science Teachers Association.

Kimball, M. (1968). Understanding the nature of science: A comparison of scientists and science teachers. *Journal of Research in Science Teaching, 5*, 110–120.

Lederman, N. G. (1992). Students and teachers conceptions of the nature of science: A review of the research. *Journal of Research in Science Teaching, 29*(4), 331–359.

National Center for Education Statistics, U.S. Department of Education. (1993). *Science framework for the 1994 national assessment of educational progress*. Washington, D.C.: U.S. Department of Education National Assessment Governing Board.

National Science Teachers Association. (1982). *Science-Technology-Society: Science education for the 1980s*. Position Paper. Washington, D.C.: Author.

National Science Teachers Association. (1992). *Scope, sequence, and coordination of secondary school science: Volume I. The content core. A guide for curriculum designers*. Washington, D.C.: Author.

Rutherford, F. J., and Ahlgren, A. (1989). *Science for all Americans*. New York: Oxford University Press.

CHAPTER 14

AN STS APPROACH ACCOMPLISHES GREATER CAREER AWARENESS

Chin-Tang Liu
Robert E. Yager

WHAT WE KNOW

Project Synthesis (Harms and Yager, 1981), a large project funded by National Science Foundation, provided a synthesis of vast quantities of information gathered by other NSF studies to determine the status of science education just prior to the 1980s. More than 3,000 pages of data from three status studies conducted by Helgeson, Blosser, and Howe (1977), Weiss (1978), Stake and Easley (1978), and the Department of Education's *The Third Assessment of Science, 1976–1977* (National Assessment of Educational Progress, 1978) were the major sources of data for Project Synthesis. Four goal clusters were identified by the project and provided a framework for synthesis efforts. In essence, these goals became the justifications for school science and the criteria for judging the appropriateness of curriculum and instructional strategies used. Perhaps establishing the four broad goal areas was the most important contribution of the project. The four goal clusters include:

1. Personal Needs. Science education should prepare individuals to utilize science for improving their own lives and for coping with an increasingly technological world.
2. Societal Issues. Science education should provide informed citizens prepared to deal responsibly with science-related societal issues.
3. Career Awareness. Science education should give all students an awareness of the nature and scope of the wide variety of science- and technology-related careers open to students of varying aptitudes and interests.
4. Academic Preparation. Science education should allow students who are likely to pursue science academically as well as professionally to acquire the academic knowledge appropriate for their needs.

When the four cluster goals of Project Synthesis are used, broader objectives for science education emerge. In the 1985-1986 National Survey of Science and Mathematics Education (Weiss, 1987), teachers were given a list of possible objectives of science education and asked how much emphasis each received in a randomly selected class. The survey showed a continuous growth of concerns about the career relevance of science. These results prompted interest in the possible effectiveness of a reform approach like STS for achieving more success with science career awareness for more people.

Such student perceptions concerning various science careers have received much attention because of the declines in student interest for pursuing science-related careers during the past five decades, especially for underrepresented women and minority students in science and mathematics. Reasons for the negative feelings and attitudes held by students toward science or the study of science may be attributed to the teaching of science rather than to the nature of the subject itself (Brush, 1983; Mason, 1983).

Based on the concerns about typical science teaching assisting with further decline in career awareness and interest regarding science, efforts such as the National Science Teachers Association's Search for Excellence in Science Education (SESE) were initiated. Basic to SESE was the establishment of specific features for excellence. In 1987, a task force chaired by Walter Smith identified criteria for Exemplary Science Teaching focusing on Career Awareness. Briefly, these programs were characterized as those which:

1. Engage students individually and cooperatively in hands-on science activities related to careers;
2. Use individual and community resources for technological updates and career information;
3. Include a range of learning strategies, levels of difficulty, and types of rewards appropriate for the diversity of students.

As studies are reviewed for ERIC there is great interest in factors that affect female and minority students choosing science and science-related careers. In this regard Baker (1987) found that females who prefer science-related careers have a more masculine perception of themselves and, to a lesser degree, a more positive role-specific self-concept. Studies by Stake and Granger (1978) and Smith and Erb (1986) assert that female students with same-sex teacher models have a higher science-career commitment. Moreover, Stake and Granger point out that the scarcity of female science teachers in the high school is a factor that discourages girls from aiming for science careers.

In order to change such attitudes and aspirations for considering science and science-related careers, many suggestions have been offered. In science classrooms, Tobin and Garnett (1987) suggest that teachers should be sensitized to gender differences concerning student participation (e.g., public interaction with the teachers) and that they should develop the necessary skills to provide equal engagement opportunities for all students. Further, a study by Hill, Pettus, and Hedin (1990) suggests that to change student perceptions about science careers, new and effective strategies must be given to instructors for the teaching of science that take into account factors that have tended to alienate girls in the past. They suggest using the social context of science as a framework within which the conceptual structure of the subject could influence the response of girls to science. Similarly, several studies also suggest that using experiences from real life, especially those that introduce humanistic and socially relevant components, will improve the situation (Erickson and Erickson, 1984; Kahle and Lakes, 1983; Kelly et al., 1984).

The STS approach has been identified by National Science Teachers Association, Project 2061: Science for All Americans and with the NSTA/NSF Project: Scope, Sequence, and Coordination (SS&C) as one with great potential for meeting the first three goal clusters advanced by Project Synthesis, namely personal needs, social issues, and career awareness.

Research Questions

Since STS is primarily an approach to instruction, it is of great interest to consider what teachers can do to improve past failures to encourage more to pursue science and technology careers. Special considerations concerning the gender effects with STS instruction are also of interest. The specific questions of this study include:

1. Are there significant differences between classes taught with STS and Textbook-oriented approaches in terms of assisting students in the attainment of more accurate and more positive perceptions concerning science careers?
2. Are there significant differences between male and female students taught with the same STS approach in terms of their attainment of more accurate and more positive student perceptions of science careers?

Methodology

Fifteen "Lead Teachers" associated with the Iowa Chautauqua Inservice Program from grades four through nine agreed to help test the effectiveness of the STS approach in terms of changing student perceptions of science careers. Lead Teachers are those who have used STS strategies successfully for two to five years and who have participated in Leadership Development Workshops for the Iowa Chautauqua Inservice Program. In this study, they adopted two different instructional strategies, namely the STS approach and a textbook-oriented approach, in two of their classrooms for one semester. Classes were randomly assigned to treatments by school administrators and counselors. Checks were made by counseling staffs to determine similarities of the two groups of students as to gender, socioeconomic balance, science interest, and general aptitude. Both sections selected for the study for a given teacher were judged to be as similar in student makeup as possible.

Fidelity of treatment was controlled with videotapes of three classes for each teacher in each teaching mode as they utilized the two instructional approaches (Liu, 1992). Videotapes were collected weekly to illustrate how time and topics were held central; the only difference between STS- and textbook-oriented classes was the teaching strategy employed. Analyses of the data revealed significant differences in the two treatments in the following ways:

1. Number of questions teachers asked during the class period;
2. Amount of information provided by teachers in a single time frame;
3. Extent of the use of student questions to drive discussions;
4. Amount of time teachers spend at the front of the classroom; and
5. Amount of time teachers spend interacting with individual students or students in cooperative groups.

In all five areas the teachers in the STS sections were significantly better than when in the Non-STS mode for points 1, 3, and 5. The opposite situation was found with respect to points 2 and 4.

Instrumentation

"Student Perceptions of Science Careers" was an instrument developed for the purpose of evaluating student perceptions concerning various science careers. These five items were chosen from those included in *The Third Assessment of Science, 1976–1977* (NAEP, 1978). The five items included in *The Iowa Assessment Handbook* (Yager, Kellerman, and Blunck 1992) in this area include:

1. Science would be an area where I would be interested to work after I graduate from school;
2. Science as a way of earning a living would be exciting for me;
3. Being a scientist requires talents only a few have;
4. Society needs to support the preparation of more scientists for the future;
5. Work as a scientist would make me rich.

The instrument was a Likert-type scale with four points for student response (i.e., 4—Very highly agree, 3—Moderately agree, 2—Mildly disagree, 1—Do not agree). The reliability of the instrument was found to range from 0.81 to 0.93 through testing and retesting a week later.

Data Collection

Pretesting was completed in September of the new academic year by the fifteen Lead Teachers for the 1989–1990 academic year. This effort was their action research project for the year. At least one semester action research effort is a required part for being a Chautauqua Lead Teacher. Posttesting of students concerning various science careers was completed in a similar manner at the end of the first semester. All pre- and posttests were administered by the instructors, then mailed to the central Chautauqua office.

Data Analysis

The data for the dependent variable, that is, student perceptions about science careers, were evaluated by analysis of covariance with treatment group (research question 1) and gender (research questions 2 and 3) as independent variables. Both theoretical and practical considerations resulted in collapsing the grade level into one group for analysis. Of prime interest was looking at the effects of instructional strategies as they correspond to gender difference for questions 1 through 3.

To analyze the data, analysis of covariance procedures were used to compute the significant differences in student perceptions of science careers. In all cases the pretest scores were used as the covariate to control for variance in the dependent variable caused by perceptions existing prior to instruction. The SYSTAT (1992) computational program (PC) was used to analyze all the statistical results.

Results

The results concerning the effects of STS on student perceptions of science careers are presented in Tables 14.1 through 14.3. The data indicate (see Table 14.1) that there are significant differences ($P < 0.001$) for all items on posttest selections for students in STS- and Textbook-oriented classroom, when using the pretest scores as the covariate. The data show that there are significant differences ($P < 0.001$) for all the 15 pairs of classes taught by 15 different

TABLE 14.1 Comparison of Students' Perceptions of Science Careers in STS and Textbook-Oriented (Non-STS) Classrooms (in Percentages)

	STS Classroom N = 364		*Non-STS Classroom N = 359*		
Choice	*Pre*	*Post*	*Pre*	*Post*	*F*
Item 1. Science Would Be an Area Where I Would Be Interested to Work after I Graduate from School					
Very Highly Agree	12.36	33.53	13.37	10.58	
Moderately Agree	25.99	52.47	32.31	38.16	
Mildly Disagree	35.71	1.92	38.00	37.60	
Do Not Agree	15.93	12.09	15.32	13.65	
					322.4*
Item 2. Science as a Way of Earning a Living Would Be Exciting for Me					
Very Highly Agree	15.38	38.74	13.09	10.31	
Moderately Agree	29.40	45.05	31.48	34.26	
Mildly Disagree	39.01	15.66	41.78	41.50	
Do Not Agree	16.21	0.55	13.65	13.93	
					755.4*
Item 3. Being a Scientist Requires Talents Only a Few Have					
Very Highly Agree	24.45	34.62	22.01	16.16	
Moderately Agree	71.15	62.64	72.98	70.75	
Mildly Disagree	4.40	2.75	5.01	12.81	
Do Not Agree	0.00	0.00	0.00	0.28	
					612.0*
Item 4. Society Needs to Support the Preparation of More Scientists for the Future					
Very Highly Agree	7.42	40.11	6.41	7.52	
Moderately Agree	37.36	53.85	38.72	36.21	
Mildly Disagree	51.65	5.77	52.09	49.58	
Do Not Agree	3.57	0.77	2.99	6.69	
					1,145.0*
Item 5. Work as a Scientist Would Make Me Rich					
Very Highly Agree	0.55	0.27	0.84	0.84	
Moderately Agree	76.10	3.85	74.93	67.97	
Mildly Disagree	23.35	83.24	24.23	30.36	
Do Not Agree	0.00	12.64	0.00	0.84	
					683.6*

Note: F = F-values of analysis of covariance (ANCOVA) in comparing posttest scores with the pretest scores as the covariate.

* Significant at the 0.001 level of confidence.

TABLE 14.2 Perceptions of Female and Male Students Concerning Science Careers after Experiencing STS Instruction

	Female (N = 187)				*Male (N = 177)*					
	Mean		*S.D.*		*Mean*		*S.D.*			
Item	*Pre*	*Post*	*Pre*	*Post*	*Pre*	*Post*	*Pre*	*Post*	*F*	*P*
1	1.88	2.50	0.81	0.85	1.97	2.61	1.01	0.95	0.01	0.92
2	1.70	2.64	0.73	0.81	1.86	2.48	0.71	1.06	53.25*	0.00
3†	2.86	1.78	0.38	0.44	2.49	1.83	0.59	0.55	97.05*	0.00
4	1.63	2.58	0.61	0.62	1.70	2.44	0.58	0.75	26.37*	0.00
5†	3.17	2.29	0.47	0.47	2.99	2.16	0.32	0.38	15.06*	0.00

Note: F, P = F- and P-values of analysis of covariance (ANCOVA) in comparing posttest scores with the pretest scores as the covariate.
* Significant at the 0.05 level of confidence
† Negative item, lower posttest scores indicate positive results

TABLE 14.3 Perceptions of Female and Male Students Concerning Science Careers after Experiencing Typical Textbook Instruction

	Female (N = 180)				*Male (N = 179)*					
	Mean		*S.D.*		*Mean*		*S.D.*			
Item	*Pre*	*Post*	*Pre*	*Post*	*Pre*	*Post*	*Pre*	*Post*	*F*	*P*
1	2.42	2.43	0.79	0.89	2.67	2.70	0.91	0.91	0.08	0.78
2	2.52	2.49	0.81	0.77	2.67	2.62	0.89	0.99	0.73	0.39
3†	1.98	1.78	0.50	0.46	1.96	1.88	0.59	0.52	5.09*	0.03
4	2.54	2.50	0.71	0.62	2.57	2.53	0.75	0.70	0.03	0.86
5†	2.34	2.28	0.52	0.48	2.99	2.18	0.48	0.40	0.77	0.38

Note: F, P = F- and P-values of analysis of covariance (ANCOVA) in comparing posttest scores with the pretest scores as the covariate.
* Significant at the 0.05 level of confidence
† Negative item, lower posttest scores indicate positive results

teachers. The STS approach has significant impact on students in developing more positive perceptions of science careers compared to students experiencing the Textbook-oriented (Non-STS) approach.

Items 3 and 5 are negative perceptions since science is "not only for the most talented," and since science may not "make somebody rich." The increasing responses under "I do not agree" category of the posttest in items 3 and 5 were found in the STS classrooms. This indicates that students develop more accurate perceptions concerning these two items (3 and 5) in STS classrooms since their perceptions become negative, which is positive in terms of affecting carrer interest more accurately.

For the gender effects, the results of statistical analyses between results for female and male students are presented in Tables 14.2 and 14.3. The data shown in Table 14.2 include the F-values and P-values for the Analyses of Covariance. On the one hand, the data indicate that there are significant differences ($P < 0.001$) between female and male students on posttest scores in the STS classrooms with the exception of item 1. The analyses use the pretest scores as the covariate. On the other hand, the data shown in Table 14.3 indicate that there is no significant difference between female and male students on posttest scores with the exception of item 3 in the textbook-oriented (Non-STS) classroom. This item ("Being a scientist requires talents only a few have") is actually a negative perception.

From these results, one can conclude that the STS approach not only enables female and male students to improve in terms of their perceptions concerning various science careers, but the approach also offers a greater impact on female students.

Discussion and Interpretation

More students in STS classrooms indicate that they would be interested in working in science fields after graduating from school than the number found in classrooms where the textbook is used regularly to determine what is taught and how it is taught. Students indicate they would be excited with science as a way of earning a living. They also feel that society needs to support the preparation of more scientists for the future. In both cases, similar perceptions were not found in textbook-dominated sections. These results indicate that STS teaching helps female and male students construct more positive perceptions about science careers.

Moreover, as we compared the female and male students in STS classrooms and in textbook-oriented classrooms respectively, we found, on one hand, that female students have more positive perceptions. Specific statements to illustrate this include: "Science as a way of earning a living would be exciting for me," and "Society needs to support the preparation of more scientists for the future." But, we found that females disagree with two perceptions, namely: "Being a scientist requires talents only a few have," and "Work as a scientist would make me rich." The level of disagreement was greater than it was for male students even though they all experienced the same instructional strategies.

It is important to note that there are no significant attitude differences between female and male students in the Non-STS classrooms. Reasons for these results may be directly or indirectly related to the ways teachers behaved in their classrooms. In STS classrooms, teachers ask students to be actively engaged in planning and learning science activities; students are expected to extend their learning to their homes and their communities. Further, they are expected to be more socially emotionally engaged in their own learning.

The findings of this study were similar to the research of Banerjee and Yager (1992). Their experiments supported the contention of Hill, Pettus, and Hedin (1990) that the use of the social context of science is an effective framework for stimulating more positive responses among female and minority students to science. Again, this finding is supported by the studies of Erickson and Erickson (1984), Kelly et al. (1984), and Kahle and Lakes (1983).

The instructional strategies utilized in the STS approach match the criteria of instruction in exemplary programs designed to improve student career awareness in science that were identified by Smith (1987). In addition, the STS approach also offers a clear advantage to student learning in various other domains, namely concept mastery, development of science process skills, application of concepts, attitude, creativity skills, and the nature of science (Liu, 1992). Today, as we try to improve science in kindergarten through twelfth-grade settings, the power of the STS approach should be noted. Certainly, this study provides strong evidence for STS as fundamental reform in science education by stimulating greater career awareness, especially among female students.

General Findings

Despite much research effort to improve student perceptions concerning science careers, such perceptions remain relatively negative as students progress through grades kindergarten through twelve. This study focuses on how the instructional approach changes student perceptions concerning science careers. A total of 15 science teachers and 722 students in grades four through nine participated in this study. Each teacher used two different instructional approaches (namely, STS and textbook-oriented) in each of two science classes. In addition to the effect of instructional approaches on student perceptions concerning science careers, the gender factor was investigated. The results indicate that students taught with the STS approach attain significantly more accurate and positive perceptions concerning science careers than students taught with the typical textbook-oriented approach. Moreover, female students taught with the STS approach attain significantly more accurate perceptions concerning science careers than male students taught with the same STS approach.

References

Baker, D. R. (1987). The influence of role-specific self-concept and self-role identity on career choices in science. *Journal of Research in Science Teaching*, *24*(8), 739–756.

Banerjee, A. C., and Yager, R. E. (1992). Improvement in student perceptions of their science teachers, the nature of science, and science careers with science-technology-society approaches. In R. E. Yager (ed.), *The status of science-technology-society reform efforts around the world, 1992 ICASE yearbook* (pp. 102–109). Hong Kong: International Council of Associations for Science Education.

Brush, L. H. (1983). Avoidance of science and stereotypes of scientists. *Journal of Research in Science Teaching, 16*(3), 237–241.

Erickson, G. L., and Erickson, L. J. (1984). Females and science achievement: Evidence, explanation, and implication. *Science Education, 68*(2), 63–89.

Harms, N. C., and Yager, R. E. (eds.). (1981). *What research says to the science teacher Volume 3*. Washington, D.C.: National Science Teachers Association.

Helgeson, S. L., Blosser, P. E., and Howe, R. W. (1977). *The status of pre-college science, mathematics, and social science education: 1955–75. Vol 1: Science education*. Columbus, Ohio: Center for Science and Mathematics Education, The Ohio State University.

Hill, O. W., Pettus, W. C., and Hedin, B. A. (1990). Three studies of factors affecting the attitudes of black and females toward the pursuit of science and science-related careers. *Journal of Research in Science Teaching, 27*(4), 289–314.

Kahle, J. B., and Lakes, M. K. (1983). The myth of equality of science classrooms. *Journal of Research in Science Teaching, 20*(2), 131–140.

Kelly, A., et al. (1984). Girls into science and technology: Final report. United Kingdom, England. ED250203.

Liu, C. (1992). *Evaluating the effectiveness of an inservice teacher education program: The Iowa Chautauqua program*. Unpublished doctoral dissertation, University of Iowa, Iowa City.

Mason, C. L. (1983). Working with the scientifically creatively gifted. *Proceedings of the Indiana Academy of Science, 93*, 375–378.

National Assessment of Educational Progress. (1978). *The third assessment of science, 1976–1977*. Denver: Author.

Smith, W. S., and Erb, T. O. (1986). Effect of women science career role models on early adolescents' attitudes toward scientists and women in science. *Journal of Research in Science Teaching, 23*(8), 667–676.

Smith, W. S. (1987). Forward: The search for excellence in science teaching and career awareness. In J. E. Penick (ed.), *Focus on excellence: Science teaching and career awareness, Volume 4, No. 1* (pp. 6–7). Washington, D.C.: National Science Teachers Association.

Stake, R. E., and Easley, J. (1978). *Case studies in science education, Volumes I and II* (Stock No. 038-000-00364). Washington, D.C.: U.S. Government Printing Office, Center for Instructional Research and Curriculum Evaluation, University of Illinois at Urbana-Champaign.

Stake, J. E., and Granger, C. R. (1978). Same-sex and opposite-sex teacher model influences on science career commitment among high school students. *Journal of Educational Psychology*, *70*(2), 180–186.

SYSTAT. (1992). SYSTAT for the Macintosh, Version 5.2 [Computer program]. Evanston, Ill.: SYSTAT, Inc. (ISBN 0-928789-11-X)

Tobin, K., and Garnett, P. (1987). Gender related differences in science activities. *Science Education*, *71*(1), 91–103.

Weiss, I. R. (1978). *Report of the 1977 national survey of science, mathematics, and social studies education: Center for educational research and evaluation*. Washington, D.C.: U.S. Government Printing Office.

Weiss, I. R. (1987). *Report of the 1985–86 national survey of science and mathematics education*. Research Triangle Park, N.C.: Center for Educational Research and Evaluation.

Yager, R. E., Kellerman, L., and Blunck, S. M. (1992). *The Iowa assessment handbook*. Iowa City: University of Iowa, Science Education Center.

PART III

What the STS Approach Demands

The STS approach requires some major changes in teachers. This, of course, includes new ways of preparing teachers.

The STS approach demands a new type of textbook. It surely demands new ways of defining scientific literacy and new ways of assessing successes.

Most current reforms (e.g., Project 2061 of the American Association for the Advancement of Science) proclaim that "less is more." This means that the number of concepts need to be reduced and perhaps reclassified as themes. Perhaps, these should be used as organizers in place of typical topics that comprise college science courses. Six themes suggested by Project 2061 illustrate this point: systems, models, constancy, patterns of change, evolution, and scale. How one selects and organizes content is a problem for STS—since the students must be involved in the selection and organization process as well. This means that general titles that promote integrated science and organized around real-world problems represent endpoints—not a highway system for all to follow day-by-day.

Another "demand" for STS is some considerable attention to technology, the connector in STS. A controversial demand is the focus on society—the second *S* of STS. But it is "society" that provides the questions and issues as well as a means of taking corrective actions after study, experimentation, and reflection.

Chapter 15

Different Goals, Different Strategies: STS Teachers Must Reflect Them

John E. Penick
Ronald J. Bonnstetter

What STS Requires

STS is an instructional approach requiring sensitivity to students, goals, and the role of the teacher. Forget one of these and you will not have an effective STS classroom environment. STS teachers tell us their goals for students who complete a full-primary and secondary-science instructional program involve far more than knowledge of science skills and concepts. Every time we ask, regardless of school or location, the goals are almost always the same and reflect much concern for student attitude and application of knowledge and little for traditional science content knowledge. We often have to suggest, "What about science knowledge?" before it is included. Even then, no one wants knowledge for its own sake; teachers always suggest "use knowledge to solve problems" or "apply their knowledge of science." Unfortunately, goals expressed by teachers do not always predict their actions in the classroom.

What We're After

STS teachers stress development of personal characteristics of learners and indicate students should appreciate and apply knowledge. Such goals tend to be relatively broad and of the sort you work toward but never fully achieve. Typical goals stated by STS teachers include students who:

1. Use their knowledge in identifying and solving problems;
2. Are more creative;
3. Take action based on evidence and knowledge;

4. Communicate science effectively;
5. Have a positive attitude toward science;
6. Know how to learn science.

Teachers with these goals want their students to learn as much science as possible. Many (Good and Brophy, 1991; Harms and Yager, 1981) have pointed out, however, that while teachers express concern with other than traditional learning, few overtly teach for and even fewer evaluate for these noncognitive goals. Some would be even stronger, saying that without these particular goals, a science curriculum lacks the essential elements that promote generalized scientific literacy (Penick, 1993; Project 2000+, 1992). This usually means understanding, appreciating, and applying science and scientific knowledge.

STS teachers systematically create new approaches to instruction and consistently remember to make the goals important and visible in the classroom. Their students know these are important personal goals. When students know their teacher's broad goals, they fully expect their teacher to model the same goals, a most powerful and persuasive situation. As a result, STS instruction is noticeably different from science teaching in general. And, since students copy their teacher's behavior to some extent (Anderson and Brewer, 1946), these classrooms are effective as well.

Several of the goals suggest considerable student activity and even independence. In addition, goals such as creativity should lead to much diversity in what students are doing in the classroom. Yet, goals are not necessarily reality. As Goodlad (1983) noted, "In our visits to (traditional) schools, we received an overwhelming impression of student passivity" (p. 554). Others (Yager and Penick, 1983) remind us that what little action students take in the typical classroom is usually directed toward more traditional and limiting goals reflecting little more than memorizing content knowledge or practicing basic skills. Overall, students are generally not even working toward, much less achieving, the goals most say we want and our students need. Much of this failure can be traced to the classroom climate established by the teacher. STS teachers cannot be typical or traditional; they must teach with strategies designed to achieve new goals.

The STS teacher is clearly aware of how to achieve these goals and then teach in ways compatible with the goals, research about learning, and the nature of science. Effective STS teachers are aware of their behavior (both desired and actual) and that of their students. As Withal (1972) put it, the professional teacher "constantly observes what he is doing and collects feedback on the impact of his behavior on the target person" (p. 332). This is conceptualized as a three part model (see Figure 15.1) where the desired goal dictates and drives the roles of both student and teacher, since each is dependent on the other.

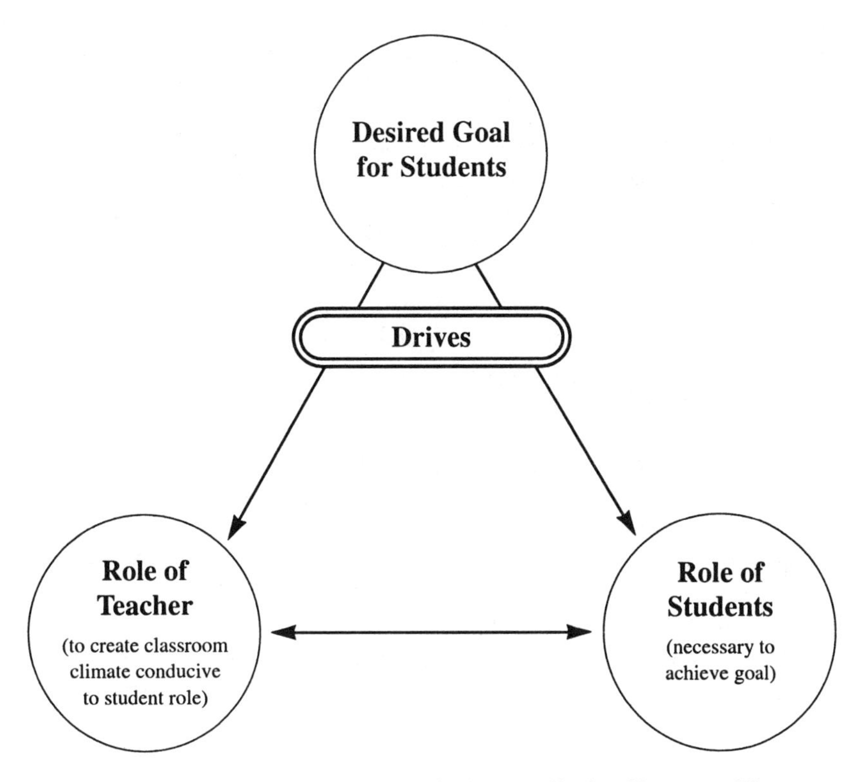

FIGURE 15.1 A Three-Part Model for Designing an Effective Classroom Climate

THE ROLE OF STUDENTS

Considering each of the six student goals, we can easily develop an image of what students should be doing (their role) in an STS class if they are consistently working toward the stated goals and constructing desired meaning, attitudes, and action. For instance, if students are learning to communicate science effectively (goal number 4) we should see individuals reading and writing, pairs or small groups conversing or interactions with the teacher. Sometimes we would expect one student to be speaking to a group or the whole class. Since communication involves comprehension as well, we would expect to see students analyzing and responding with questions, statements, action, and feedback. And, as this is in the context of an STS science class, the subjects under discussion would be science or science-related. Even for those students initially reading or writing, we would hope, eventually, to see them use what they read or write to synthesize ideas and communicate with others. Equally important, we would see the teacher overtly teaching students how to communicate and modeling effective communication.

If education is for life and living, in effective and ambitious STS classrooms we would expect to see students carry the activity beyond the classroom walls to where students live. In working toward the goal of "effective communication" students should interact with outsiders, including politicians, leaders, scientists, teachers, and other students. Students should be writing letters, making calls and speeches, working with their local newspapers, not as academic exercises, but as real-world action with a purpose.

An STS teacher encourages such interaction, communication, and conversation and recognizes that each of the other goals requires an equally appropriate and carefully designed classroom climate, including the physical setting and the role of the teacher. In addition, since each goal implies overt action on the part of the student, the teacher must be prepared to encourage, support, and promote desired student initiatives and action. Teachers must recognize the difference between appropriate and inappropriate action on the part of students and themselves and modify their behavior and the classroom climate to be compatible with the student goals.

Encouraging intellectual freedom to stimulate critical thinking, creativity, and communication while restricting social freedom to that deemed necessary or desirable requires a teacher with a clear image of the desired classroom climate, a rationale to define and defend it, and the competence to create it.

The STS Classroom Climate

Through interaction, common experiences, and time individual students eventually become a group. Parallel to this dynamic, social process, students develop structures of perceiving and processing that are common to most or even all of them. This element of the cognitive representation shared by the class members is called classroom climate. Classroom climate is a useful construct in predicting achievement and satisfaction for students (Raviv et al., 1990).

An STS teacher must establish a climate where learning proceeds in the desired direction. Carl Rogers (1969) noted that the teacher has much to do with setting the climate by eliciting and clarifying the purposes of the individuals and the group. In doing so, the teacher often relies on the desires of students to achieve purposes that have personal meaning and significance. These desires form a strong motivational force and affect learning outcomes.

Students in high-achieving classes perceive more cooperation between teachers and students, more individualization, and more understanding than in low-achieving classes (Dressman, 1982). And, in classes with a favorable climate, students have a higher self-concept of ability and attribute their successes to internal rather than external factors.

Unfortunately, as Withall pointed out in 1972, most teachers are unaware of their behavior and its impact on learners. Good and Brophy (1991) spent most of a chapter discussing classroom problems that arise through lack of teacher awareness. They document that teachers dominate communication, overuse factual questions, do little to motivate, and, in general, neglect to emphasize meaning. Tobias (1990) made the same observations about university classes. These teachers are not mean or insensitive; rather they are not trained to note the subtleties and nuance of their actions and remarks. The STS teacher is sensitive, designing a classroom to meet explicit goals of instruction.

Rather than merely saying, *"We are having a discussion"*—an STS teacher might say: *"In a discussion I specifically ask questions that have multiple possible answers and require little prior knowledge. Then, after asking the question, I wait intently and quietly until I have a response. When a student responds, I never, never evaluate the response. I don't say 'You are right' or 'That is wrong.' Instead I ask the student to elaborate or clarify. Or I may ask an extension of the same question or for evidence to support an idea. Then, rather than continuing with my own line of thought, I ask others to comment, again using much wait time."* This teacher will be viewed as more patient and will encourage multiple responses. Since the teacher does not do most of the talking (Dillon, 1981) or know the students' answers in advance, the teacher is learning also (Rogers, 1969), a key factor in exciting STS classrooms. This is a developmental or constructivist classroom. The effective STS teacher actively encourages and wants critical thinking and creativity.

The STS teacher creates a classroom climate which organizes, stimulates, and makes readily available a wide range of resources and ideas and actions that encourage students to achieve the desired roles and goals.

The Intellectual Setting

While any teacher can and does create a classroom climate, only those who set about creating a climate with definitive intellectual attributes for student success would we call a professional STS teacher. To be a professional who can create a specific classroom climate on demand requires a teacher with a solid and cohesive rationale for teaching. Such a rationale includes the desired role of the students, including the goals, an understanding of social versus intellectual freedom, and the type of action they wish students to achieve. For a rationale to be truly useful and professional, it should be based on the latest research evidence about how students learn, the effect of teacher behaviors on students, and should describe a classroom that is compatible with the nature of science. Outstanding teachers also recognize the desirability of congruence in student and teacher roles.

The STS teacher with a rationale can describe a desired classroom state and possesses the competence to put it into action. And, between rationale and action is adequate knowledge of both science and teaching to translate and interpret the appropriate research into useful classroom behaviors, attitudes, and overall classroom environment. This interpretation and translation requires knowledge, time, experience, and thoughtfulness and is often facilitated by interactions with peers.

In the STS setting where students achieve the goals previously stated, students must have the intellectual freedom to, as Percy Bridgeman said, "Do their damnedest with their minds, no holds barred." If students are to achieve these goals, then the teacher must teach for them overtly and both teacher and student must recognize their value. To do this the teacher must make visible their desirability and see that they permeate all classroom activities and discussions. This is the essence of an STS teacher. In the effective classroom teachers do make a difference. Several studies of exemplary STS teachers (Myers, 1988; Yager, Tamir, and Mackinnu, 1991; Penick, 1992) documented nine generalizations about the role of such teachers.

Effective and Successful STS Teachers

Provide a Stimulating, Accepting Environment

Effective STS teachers provide an environment rich in human and material resources combined with numerous opportunities to initiate, study, and take action in the laboratory, library, or with others by scheduling time for individual as well as group work and by expecting action, not just words or reading. Teachers encourage student decision-making and action by not being content with mere knowledge and reports. Teachers ask for evidence and clarification, then action on ideas. Students can and will go a step further, beyond mere knowledge, if their teachers expect, encourage, and reward it.

Have High Expectations of Themselves and Students

STS teachers expect students to go a step further, well beyond the norm, and to change behaviors, knowledge, and attitude. Change may be in learning, or attitude, or it may be resolution of a real problem. Teachers have different expectations as well, recognizing that students have varying potential and interests. They expect all students to achieve personal excellence through quality performance. These high expectations carry over to the teacher as evidenced by their continual push to do more, gain more involvement, and see more resolution. They are never content with the status quo. Teachers with high expectations also see their role as more than a teacher of thirty students. They see their role as a larger one in a community context and as a model of active inquiry.

Are Models of Active Inquiry

Effective teachers read and explore widely, bringing this reading, action, and new knowledge visibly into the classroom. Students know these teachers are open to wonder, new ideas, and innovations. Exemplary teachers bring inquiry alive by presenting themselves as students, eager and willing to learn new ideas, skills, and actions. By bringing new ideas to the class, by questioning and being open-minded, these teachers are clearly models to be followed. They also model the thoughtful and rational approach to problem identification and resolution by being interested, seeking information, asking for evidence, and by taking action themselves. In essence, they are making their thinking, feelings, and approaches to learning highly visible.

Expect Students to Question Facts, Teachers, and Knowledge

Teachers who make their STS approach visible have students who more easily see that science involves a degree of skepticism, as Richard Feynman put it, "a belief in the ignorance of experts." Not an assumption of being wrong, such a belief is a quest for evidence to support a fact, idea, or position. Teachers and students who have questions, seek answers, and who can provide evidence may well become active citizens with the same attributes. Certainly, these are the core of science literacy.

Stress Science Literacy and Apply Knowledge

Science literacy develops as we demand a rational and independent approach to science. Students are expected to seek, to question, to explain, and to apply their knowledge. STS teachers are not content with students just knowing words or skills; they insist that words be used to justify, defend, or clarify larger concepts or actions. For many students, they must apply their knowledge before it truly becomes a part of them. Effective STS teachers are never satisfied with content knowledge alone. Calculating vectors in a pulley system does not have as much potential and is not as problematic for most students as does actually rigging and using a block and tackle. The same is true with societal applications where students seeking justification for a new community water treatment plant view their learning as vastly different from reading a standard text on the subject. In both instances, they can see their knowledge as it is being used, often outside the classroom.

Do Not View the Classroom Walls as a Boundary

Students and learning must go beyond the classroom. Any learning that is real must, by definition, continue with the students into the rest of their lives and world. Obviously, effective teachers see that both their students and their ideas have many opportunities to go beyond the classroom, via field trips, visits to resource persons and facilities, and using outside materials such as books, papers, and ideas. At the same time, teachers invite community experts in, offering new insights and perspective on existing problems and often raising new ones. Teachers also encourage going beyond by stressing issues and ideas from outside the school.

Are Flexible in Their Time, Schedule, and Curriculum

Students often find new ideas to pursue, old issues get sidetracked, and the teacher may lose some control. But an effective STS teacher takes advantage of this, changing as opportunity presents itself. Students learn, but not always as the teacher anticipates nor on the same schedule. The stimulation, excitement, and involvement lead to many new avenues of motivation and learning—for the teacher as well as the student.

Put in Far More Than the Minimal Time

We find that outstanding STS teachers worry about time but focus that worry on finding more time to devote to their classes and ideas. They tell us what time they have is fleeting, the days are not long enough to contain all they want students to know and be involved with. As a result, they put in hours after school and on weekends, striving to make their teaching all it can be.

Teachers Do Make a Difference

Although most students do not like science (Mullis and Jenkins, 1988), students from STS classrooms where teachers exhibit these nine generalizations feel differently (see Table 15.1 for some specific suggestions). They like science, science teachers, and science teaching. They feel science is useful, now and in the future. Effective teachers cause students to see uses for their science and give them confidence in their own ability to use and be successful with science. And, these teachers are eager for Monday morning (Bonnstetter, 1984; Krajcik and Penick, 1989). These are the teachers we wished we had and whom we continue to want for our children. Effective teachers do make a difference and they can add even more if they have and use a well-researched and valid rationale for teaching science. The STS approach to teaching is both valid and effective. But, it requires true dedication, study, and practice to implement. The effort, however, is well worth it as students and teachers come alive in the STS classroom.

TABLE 15.1 How to Be a More Effective STS Teacher

1. Provide a safe, stimulating, accepting environment: The classroom must be rich with books, magazines, newspapers, art, equipment, and artifacts. Equipment should be readily available for spontaneous use (often students don't know they need something unless they see it).

   The teacher must stimulate with provocative and insightful questions and statements then, to encourage students, allow sufficient wait time and accept (with no evaluation) all responses. Rather than "Good" or "No, I don't think so," nod, say "Okay, what else," or get additional students to respond. Students must feel safe if they are to risk something new. It's never really safe if someone is evaluating you. Evaluation should come later as students look at causes and consequences of ideas or proposed activity. And, they need to learn how to do this evaluation themselves.
2. Have high expectations of yourself and students: Don't be content with last year or the norm; always go at least one step further. Let students know you have faith in them that they can succeed (although it may be hard work). When they produce a report, go further and have them find out who might be interested in it. Send it to your local newspaper. Rather than grading on a curve, establish levels of competence or seek evaluations that reflect more than an arbitrary level.
3. Be a model of active inquiry: Bring in questions from your own observation, reading, and conversation. Take action to find out new facts, concepts, theories, resources and information. Be a learner. Let your students know you don't already know everything. Show your curiosity.
4. Get students to question facts, teachers, and knowledge: First it must be safe. Then raise questions yourself, asking always for evidence supporting assertions or ideas. Put yourself in positions where students can successfully challenge you. Be willing to learn alongside them (see 3 above). Play "devil's advocate," raising issues and questions even if you know them to be false or irrelevant.
5. Stress scientific literacy and get students to apply knowledge: Students who see you and themselves engaged as numbers 1–4 describe will see and experience science as an active process of thinking and doing. As a result, they will come to rely on and expect their science to be useful personally. Use real-world problems and encourage your students to find other examples in their own lives. Provide time to discuss applications and evaluate these applications as well. Visibly apply knowledge yourself as you expect them to do the same.
6. Do not view classroom walls as boundaries: Information as well as students flow in and out; take advantage of it. Reward students who bring ideas and information in and take solutions out. Do the same yourself by bringing in people and materials as resources—schedule group field trips and encourage individual students to make their own field trips. Teach students to use the library, write letters, and telephone sources. Get parents involved in the classroom and out of it as well.

(continued)

TABLE 15.1 *(continued)*

7. Be flexible: Lesson plans are not religious documents. Feel free to postpone, ignore, change, or delete. Watch for "teachable moments" when you and your students can move ahead as learners. Use current events as they happen and help students to be "investigative reporters." A flexible student or teacher follows leads as they happen, not when scheduled.
8. Put in more than minimal time: Then, be more efficient; get students to do more of the work. We know of many classes where the students really do the lesson planning while the teacher plans the learning environment. But it still takes time to develop your organizational plan and to have materials ready.
9. Make a difference: Trust yourself and your students. Look inward and ask, "Is this best for my students and goals?" Study exemplars and most of all, study your own teaching, making changes for the better. Your students will notice, appreciate your efforts, and be rewarded with a better teacher and classroom climate, both of which will enhance their own success.

REFERENCES

Anderson, H. H., and Brewer, J. E. (1946). Studies of classroom personalities, II: Effects of teachers' dominative and integrative contacts on children's classroom behavior. *Applied Psychology Monographs*, No. 8.

Bonnstetter, R. J. (1984). *Characteristics of teacher associated with an exemplary program compared with science teachers in general.* Unpublished doctoral dissertation, University of Iowa, Iowa City.

Dillon, J. T. (1981). Duration of response to teacher questions and statements. *Contemporary Educational Psychology*, *6*, 1–11.

Dressman, H. (1982). Classroom climate: Contributions from a European country. *Studies in Educational Evaluation*, *8*, 53–64.

Good, J. E., and Brophy, T. L. (1991). *Looking in classrooms* (5th ed.). New York: HarperCollins.

Goodlad, J. (1983). *A place called school.* New York: McGraw Hill.

Harms, N., and Yager, R. E. (1981). *What research says to the science teacher, Volume 3.* Washington, D.C.: National Science Teachers Association.

Krajcik, J. S., and Penick, J. E. (1989). Evaluation of a model science teacher education program. *Journal of Research in Science Teaching*, *26*(9), 795–810.

Myers, L. H. (1988). *Analysis of student outcomes in ninth grade physical science taught with a science-technology-society focus versus one taught with a textbook orientation.* Unpublished doctoral dissertation, The University of Iowa, Iowa City.

Penick, J. E. (1992). STS instruction enhances student creativity. In R. E. Yager (ed.), *The status of science-technology-society reform efforts around the world* (pp. 81–86). Hong Kong: International Council of Associations for Science Education.

Penick, J. E. (1993). Scientific literacy: An annotated bibliography. ICASE/UNESCO, Paris.

Project 2000+. (1992). Toward scientific and technological literacy for all: A world conference for 1993. UNESCO: Paris, France.

Raviv, A., Raviv, A, and Reisel, E. (1990). Teachers and students: Two different perspectives? Measuring social climate in the classroom. *American Educational Research Journal, 27*(1), 141–157.

Rogers, C. (1969). *Freedom to learn.* New York: Merrill.

Tobias, S. (1990). They're not dumb, They're different: Stalking the second tier. Tucson: Research Corporation.

Withall, J. (1972). Research in systematic observation in the classroom and its relevance to teachers. *Journal of Teacher Education, 23*(3), 330–332.

Yager, R. E., and Penick, J. E. (1983). School science in crisis. *Curriculum Review, 22*(3), 67–70.

Yager, R. E., Tamir, P., and Mackinnu (1991). The effect of an STS approach on achievement and attitudes of students in grades four through nine. Manuscript submitted for publication.

CHAPTER 16

TEXTBOOKS WITH SPECIAL QUALITIES FOR STS

Robert E. Yager
Betty Chiang-Soong

CHANGING DIRECTIONS FOR SCIENCE CONTENT

The Emerging National Standards for school science are leading the profession to new goals. Until now the goals have been derived from the knowledge structure of the various disciplines of science; they have arisen from scholarly research efforts of scientists. As renewed interest for a science education for all kindergarten through twelfth-grade learners is advanced, goals are now derived from analysis of the ways and means that science and technology influence society and human affairs. The NRC (1994) National Standards indicate this broader view of science *content*. Eight areas are offered as defining science content, namely Science as Inquiry, Physical Science, Life Science, Earth and Space Science, Unifying Concepts and Processes, Science and Technology, Science and Societal Challenges, and History and Nature of Science. As agreement on these broader goals for science becomes more universal, the important scientific and technological systems in cultural, societal, and individual contexts is emphasized as opposed to discipline specific concepts judged important for all to know.

The subject matter for the science curriculum has too often merely been selected by professionals from science disciplines who look at the current theories, ideas, and approaches of their disciplines. This characterized the impressive and comprehensive efforts of the 1960s. Now, however, the content must be selected in terms of its importance for conceptualizing the role of science and technology in civic affairs, for promoting citizenry responsibility, and improving the quality of life.

The cognitive processes, on the one hand, have been emphasized as analytical, deductive, theoretical, and value-free. They have been taught as pre-planned exercises that present an idealized picture of the way scientists operate. They focus on techniques, accuracy, objectivity, deductive reasoning, and rediscovery. The new goals, on the other hand, require science to be considered in a social context, which means holistic views, complexity, qualitative concerns, issues that are value-laden, and ending with decision-making and personal action. These new goals require systematic thinking and actions. Explanations are not the product of deductive reasoning; instead they are advanced as possibilities and useful outcomes which fit more easily when teaching/learning goals are offered as a part of a social system. The new goals emphasize information processing and the use of knowledge in everyday situations.

Such a view and definition for school science are basic to STS and suggest fundamental changes in the characteristics and function of textbooks in the program. Formerly, textbooks provided a source of information concerning scientific advances and the general information identified as important for the pursuit of further study of science. STS moves the primary function of textbooks to but one source for valid information concerning current problems. Voelker (1982) summarizes this view with his analysis of the development of scientific literacy in the early 1980s:

> . . . if we want a science program that is truly responsive and responsible to the citizen in a scientifically and technologically oriented society, we must elevate current and future citizen concerns. We cannot assume that curricula which emphasize traditional cognitive knowledge and an understanding of the scientific process will lead to an understanding of the science-related issues confronting society. Neither can we assume that such traditional curricula will assist our student-citizens in applying their scientific knowledge and processes to these issues. Some sacred cows of the science curriculum must be eliminated. But the short-term trauma this sacrifice may elicit may be replaced by a long-term gain for all citizens. (p. 79)

Changing textbooks to reflect the newer goals, newer focus for the curriculum, and improved instructional strategies could occur relatively easily. However, such changes demand that science teachers use textbooks differently. In 1977 the Educational Products Information Exchange (EPIE) Institute indicated that the experiences for 95 percent of the students are textbook-based. The NSF Status Studies released at the same time (Helgeson, Blosser, and Howe, 1977; Stake and Easley, 1978; Weiss, 1978) verified the almost universal use of textbooks for determining science content, activities, and evaluation for most students enrolled in kindergarten through twelfth grade (Harms and Yager, 1981). Teachers feel dependent on textbooks. Stake and Easley (1978) reported that many were dissatisfied with the textbooks in use but felt powerless to do anything but use them—and on a daily basis.

Although some recent surveys have suggested changes in attitude (both for teachers and members of the general public) (Pogge and Yager, 1987) concerning the centrality of the textbook in the science classroom, Weiss's 1987 national survey contained some disconcerting information. She reported that three-fourths of all kindergarten through twelfth-grade science teachers indicate little or no problem with the nature of current science textbooks. This no doubt indicates that by far the majority of teachers continue to view science as information packaged in different ways at different grade levels for students to consider and to commit to memory. Testing focuses on the degree that students can show that they have acquired the information long enough to succeed with quizzes and examinations. These factors from the real-world illustrate the problem with moves to STS.

If three-fourths of the teachers see no problem with science textbooks, there is no great motivation for publishers to alter their materials. If 75 percent of the market is generally pleased with a product, why worry about the other 25 percent who no doubt have differing views on what improved textbooks would be like? The problem may not be what features idealized textbooks would have—but how to get significant numbers of schools and teachers interested in procuring them. Surely publishers will produce whatever market analysts report to be in demand. They are not in business to lead the profession—and certainly not to produce materials with marginal chances for sales.

The improvement of science textbooks will not occur until more teachers and other educational leaders demand them. To demand changes will require greater awareness of the problem with existing materials and greater awareness of and agreement with the new goals for a science education geared to producing a more literate citizenry with respect to science and technology, and a citizenry prepared to work for general improvements of humankind. Moves to STS will surely hasten the demands for change.

Linus Pauling (1983), Stephen J. Gould (1988), and Diane B. Paul (1987) have recently offered stinging analyses of the problem with science textbooks. Such attacks need to be circulated to teacher groups; inservice workshops should be conducted to emphasize the problems. Also, more time and attention are needed to analyze the emerging description of a school science that is more appropriate and meaningful for all students. Such efforts could produce more demand for the kind of science textbooks for which many yearn. As evidence mounts that textbooks with different features are in significant demand, they will become a reality. Every publisher is interested in a better product, especially if there is hard evidence that the better product will sell.

In 1993 Chiang-Soong reported on an extensive study of the eleven most-used science textbooks in secondary schools in the United States, grades seven through twelve. She included one of the NSF-supported books from the 1960s in each of the groupings, namely grades seven through nine and ten through twelve. Chiang-Soong used the topics suggested by Harms (1977) as an indication of an STS focus. Since only the textbooks were analyzed, there is no information as to *how* they were used. As indicated earlier in this monograph, the weakness of STS exists concerning the "how" science is cast as opposed to the information outlined in the curriculum or illustrated by text material.

Chiang-Soong studied the books since the evidence is so pervasive that most teachers use the textbook on a daily basis. The topics Piel (1981) used to define STS include: energy, population, human engineering, environmental quality and utilization of natural resources, space research and national defense, sociology of science, and effects of technological development. Chiang-Soong examined every page and determined the total space devoted to these STS topics.

Chiang-Soong's results indicated that there was more STS included in middle/junior high texts—on average 6.3 percent of the total included. At the high school level the amount of coverage was limited to less than 3 percent. Assuming that STS illustrates the new reform goals, it is apparent that changes are needed. Either teachers could reverse their dependence on textbooks—perhaps even abandoning the crutch altogether. Or they could demand textbooks that would assist them in making changes that reform demands.

This chapter is offered assuming that adequate demand for improving textbooks will develop and that the improved textbooks will coincide with the new goals for school science. Greater consensus concerning these goals and more moves to STS will create the demand.

Perhaps the most clear-cut statement of goals for school science was stated by Norris Harms in his proposal to NSF for the funding of Project Synthesis. These goal areas have been offered as justification for kindergarten through twelfth-grade science for all students. The goal areas are: (1) science for meeting personal needs of students; (2) science for preparing students to deal with current societal issues; (3) science for making career choices; and (4) science as preparation for more in-depth study of science at the next academic level. Analysis of current textbooks reveals that there is less than 10 percent attention for goals 1, 2, and 3. At least 90 percent of the treatment in existing books serves only to sample concepts across disciplines with the promise that such a consideration will be useful to the student enrolled and that the various disciplines of science have been fairly and equally considered.

Recently Mitman et al. (1987) reported that science teachers rarely, if ever, make explicit reference to the societal, reasoning, historical, or attitudinal implication of their courses. The textbooks in use are devoid of material to support instruction and student learning in science literacy dimensions. Soong's (Chiang-Soong and Yager, 1993) recent analysis of the eleven most used science textbooks in the secondary schools reveals that less than 5 percent of the material included responds to any goal other than the presentation of information—usually in the context of a signal discipline.

Assuming that one wants to move toward courses/programs that meet the new goals, and illustrate STS, what would the textbooks be like? Assuming that the preponderance of attention to presenting central concepts characterizing the discipline as viewed by scientists is the problem, what should be added to an appropriate textbook? What should be omitted? How should such textbooks be organized?

Broaden the Definition of Science

Science is more than the current explanations and understandings held by scientists. It is such a unidimensional view that creates major problems with students understanding of science; it contributes to mass scientific illiteracy for most of our population.

Science impacts the whole of society; and society impacts science. This interaction should be a major component of a textbook designed to introduce all youth to the ways of science.

The use of scientific knowledge for producing a variety of technologies and considering the effects of them on human society are other important dimensions. During the 1960s most were willing to ignore technology while emphasizing "science as it is known to scientists." Technology was ignored and/or relegated to vocational areas or physical education/health. It is interesting to note in retrospect how inappropriate these actions were since the applications of science produce motivation and offer the best explanation for the ultimate value of science—something that students frequently ask when considering the great quantity of science found in textbooks.

The reasoning process of science is another important dimension that should be a prominent part of a quality science textbook. Students must be encouraged to practice the skills characteristic of science. However, this treatment should not be overglamorized, thereby making science seem to be something practiced by superhumans. The reasoning process should be approached as something everyone does, can do, and can do better. The materials should show how reasoning applies to all intellectual activities—not just scientific exploration. It should include problem formulation, data gathering, inference, interpretation, and generalizing.

Still another important dimension is the historical one. Science needs to be presented as a continuing series of self-correcting human endeavors. It is not the discovery of preexisting facts. Instead it is a series of evolving contributions of a variety of peoples over time. Quality science textbooks must include information about social communications and historical settings that limit views of the universe and affect the formulation of new theories and explanations. The textbooks should illustrate how scientists as individuals enliven and impact the progress of science itself.

UTILIZE AND EXPAND STUDENT CURIOSITY

Good textbooks should motivate students; they should build on their curiosities and encourage even more questions and activities. Someone once said the real subject matter of science is:

1. What we know that we don't know;
2. What we know that isn't so;
3. What we don't even know that we don't know.

Curiosity means questions; it means wondering; it means interest. Curiosity is the first basic ingredient of science and, when encouraged and enlarged, more positive student attitudes develop. A quality textbook in science will take advantage of student curiosity, extend it, use it as bridge to a variety of considerations, activities, and uses.

Helping students develop and intensify curiosity means that they are actively involved with their own learning. They are encouraged to be directly involved and, as such, they practice following up on information gathering and are encouraged to deepen interests. There are actions required for the scientifically literate.

The writing style of science textbooks should be personal and exciting. Too often the textbooks emphasize "drill and practice" techniques when the writing is bland and voiceless—often like an encyclopedia. The writing should stimulate and catch the interest of the reader. Often attempts to lower reading level and to reduce the complexity of concepts result in misinformation and loss of life and interest.

Using more primary source materials or referring students to such sources can increase student interest—while also encouraging students to operate outside the textbook as the primary information source, inspiration, and filter. More such materials could be included in the textbook at the expense of more study skill exercises and questions.

Science textbooks should focus on more issues and problems that are inherently interesting to more students. Science is then cast as a more important human endeavor and something that is current and important to the current condition.

Student curiosity could be enforced by limiting the sheer size and bulk of the textbook. Textbooks have grown in size—some more than doubling the number of pages over earlier editions. Revision often means addition. Textbooks should provide a suggested framework that has no real boundaries. Unfortunately, the textbook itself becomes the boundary—with neither teacher nor student escaping into it. Often there is frustration for both since the material included is so great that completing "the book" in a single year is a physical impossibility. When reviewing a good textbook, one should remember that it need not—indeed should not—be a record day-by-day unit-by-unit of the material and the activities that will comprise the course.

Imply Appropriate Instructional Strategies

Quality textbooks will enhance instructional strategies that characterize the scientific enterprise. This suggests that word lists for the sake of word lists, quiz-type questions, inclusion of information for its own sake, verification-type activities should not be included. Commonly used check lists for textbook selection include many items—presumably desirable traits—that oppose appropriate strategies—that is, ones that cast science primarily as discreet information (verbiage) to be committed to memory.

The information included and the activities suggested should be related to previous student experiences, previous courses, and to the real world. The information and the activities must be connected to the previous experiences of students and provide a multiple direction for students to pursue. There are usually many ways to reach the same endpoint; a good textbook will provide examples and encouragement for taking a variety of pathways. Such variation is a good way to deal with individual differences and to promote cooperative learning.

A good science textbook (one useful for STS) will frequently offer comparisons and contrasts, causes and effects, pro and con debates, past versus current interpretation of evidence. It will provide opportunities for further data gathering, opposing interpretation, and actions. It will include decision-making strategies, opportunities, and practice. A good science textbook will stress the importance of communication and cooperation among scientists and people in general.

An exemplary science textbook will include frequent opportunities for students to do science. This includes all of the widely accepted processes of science as well as the more general actions that are valid responses to student curiosity. The difference between fact and theory, observation and explanation, cause and effect, assumption and evidence will be exemplified in terms of action, situations, and analyses of past situations.

A good science textbook will focus on information and questions that are related to the daily living of the student. The presentation and format will be personal—not impersonal—and relevant and meaningful—not tied to anything that the student knows or has experienced.

ENCOURAGE USE BEYOND A GIVEN COURSE

An exemplary science textbook will be a guide for student learning; this learning must be designed to affect a student's thinking and actions beyond the course. Too often the end point for a science course is demonstration that the information presented (in class and from the textbook) has been acquired by most students to some acceptable degree.

One way of encouraging use beyond a given course is considering how current advancements and contributions are affecting society today. A focus on current problems is an excellent way to accomplish this. However, the inclusion of specific issues in a standard textbook is difficult. It probably means referring students to newspapers and periodicals for the identification of such issues. Perhaps using publications such as Lester Brown's *State of the World 1994* book or the new periodical *World Watch* from Worldwatch Institute as major sources for ideas, issues, and potential actions.

A good text will include information and examples of how science and technology have been used to improve lives and, at times, caused unforeseen problems. Such views of the past can help with the present and even serve as a source for ideas for identifying and working on some current problems as a significant part of the specific science class.

A good science textbook will consider the relationship between people and their environment; it will promote awareness and responsibility toward that environment. It will consider the finite amount of natural resources and emphasize the need for wiser management.

Science textbooks must include more information about the vast array of careers related to science and technology. Some have estimated that as many as 90 percent of all students enrolled in schools today will have careers dependent on and/or related to science and technology. A science course that does not include a major component concerned with career awareness is not serving students well. And science textbooks must include significant information in this realm. It may be the most meaningful and for many the most important facet of their study of science.

EMPHASIZE CONFLICT AND APPROACH APPROPRIATE FOR THE UNDERREPRESENTED AND UNDERSERVED

Textbooks should not minimize issues related to the underrepresented and underserved; nor should they move toward ill-considered efforts to resolve these significant problems. Surely a vast resource is being lost in our society and for productive work as professional scientists and engineers when women, minorities, and disabled persons are not encouraged to pursue their interests in these fields. Textbooks and teachers need to focus on the issues and add positively to resolution of such problems. Special attention to such groups and how they could be better served can be a major factor in mounting specific projects in schools.

Too often in attempts to cover the textbooks, teachers give little attention to the problems of underserved groups. For that reason the textbooks can be an important factor in assisting everyone with greater awareness, resolution to act, and sensitivity.

Generally, science textbooks must interest students; this is more important than comprehensive coverage of topics. They must provide the students with an accurate, clear, and meaningful picture of science; they must build on their past conceptualizations and encourage continued growth and development. The treatment must be impressive enough that the experience will be memorable and serve students in an ever-increasing quest for more understanding. A good textbook will take advantage of natural student curiosity and encourage even more.

Any time a science textbook can be shown to stimulate student curiosity—perhaps to build on that which already exists in all people—we will have an exemplary textbook and one appropriate for STS. Certainly, it is a challenge to an author to include student explanations for the objects and events encountered during explorations. But examples can be given, suggestions made. A good science textbook must be open-ended in major ways—not just a listing of follow-up activities or optional excursions included at the very end of a chapter after vocabulary lists and study questions. An exemplary science textbook will engage students and provide ideas for more direct experiences—probably the only way that real science is learned.

More than fifty years ago John Dewey showed us clearly the way all humans really learned. His ideas and observations provided the rationale for planned reforms for another age. We seem to forget too easily the lessons of the greats of the past. Some of these lessons should be relearned if science textbooks are developed to assist us all with realizing the new goals for school science that are so exciting to many. However, we must find ways to get more professionals to adopt the goals. This is an imperative before we can expect many textbooks with the necessary features. Without them most students in schools today will be condemned to a kind of experience with science that we know to be detrimental. We must find a way to make school science an experience with real education, that is, the drawing out of meaning from students. This is in contrast to the "pouring in" model that exists as common practice in so many schools and is promulgated with standard science textbooks. These changes are required if STS is to exemplify the reforms we seek.

REFERENCES

Brown, L. R., Durning, A. Flavin, C., French, H. Lenssen, N., Lowe, M., Misch, A., Postel, S., Renner, M., Starke, L., Weber, P., and Young, J. (1994). *State of the world 1994: A Worldwatch Institute report on progress toward a sustainable society*. New York: W. W. Norton.

Brown, L. R. (ed.). (1988–1995). *World Watch* [all issues].

Chiang-Soong, B., and Yager, R. E. (1993). The inclusion of STS material in the most frequently used secondary science textbooks in the U.S. *Journal of Research in Science Teaching*, *30*(4), 339–349.

Gould, S. J. (1988). The case of the creeping fox terrier clone. *Natural History*, *97*(1), 16–24.

Harms, N. C. (1977). Project Synthesis: An interpretive consolidation of research identifying needs in natural science education. (A proposal prepared for the National Science Foundation.) Boulder: University of Colorado.

Harms, N. C., and Yager, R. E. (eds.). (1981). *What research says to the science teacher*, Vol. 3. Washington, D.C.: National Science Teachers Association (No. 471-14476).

Helgeson, S. L., Blosser, P. E., and Howe, R. W. (1977). *The status of pre-college science, mathematics, and social science education: 1955–75* (Stock No. 038-000-00362-3). Washington, D.C.: U.S. Government Printing Office, The Center for Science and Mathematics Education, The Ohio State University, Columbus, Ohio.

Mitman, A. L., Mergendoller, J. R., Marchman, V. A., and Packer, M. J. (1987). Instruction addressing the components of scientific literacy and its relation to student outcomes. *American Educational Research Journal*, *24*(4), 611–633.

National Research Council. (1994, January 10). *National science education standards: State focused review inquiry* (draft "headline" summary). Washington, D.C.: Author.

Paul, D. B. (1987, May/June). The nine lives of discredited data. *The Sciences*, 26–30.

Pauling, L. (1983). Throwing the book at elementary chemistry. *The Science Teacher*, *50*(6), 25–29.

Piel, E. J. (1981). Interaction of science, technology, and society in secondary school. In N. C. Harms and R. E. Yager (eds.), *What research says to the science teacher*, Vol. 3 (pp. 94–112). Washington, D.C.: National Science Teachers Association.

Pogge, A. F., and Yager, R. E. (1987). Citizen groups' perceived importance of the major goals for school science. *Science Education*, *71*(2), 221–227.

Simpson, G. G. (1963). Biology, the nature of science. *Science, 139*(3550), 81–88.

Stake, R. E., and Easley, J. (1978). *Case studies in science education*, Vols. I and II (Stock No. 038-000-00364). Washington, D.C.: U.S. Government Printing Office, Center for Instructional Research and Curriculum Evaluation, University of Illinois at Urbana-Champaign.

Voelker, A. M. (1982). The development of an attentive public for science: Implications for science teaching. In R. E. Yager (ed.), *What research says to the science teacher*, Vol. 4 (pp. 65–79). Washington, D.C.: National Science Teachers Association.

Weiss, I. R. (1978). *Report of the 1977 survey of science, mathematics, and social studies education*. Washington, D.C.: U.S. Government Printing Office, Center for Educational Research and Evaluation, Research Triangle Park, N.C.

Weiss, I. R. (1987). *Report of the 1985–86 national survey of science and mathematics education* (Report RTI/2938/00-FR, for the National Science Foundation). Research Triangle Park, N.C.: Research Triangle Institute.

Chapter 17

Scientific Literacy for Effective Citizenship

Jon D. Miller

Major Problems in Today's Society

Most citizens of modern industrial societies live in an age of science and technology. They live in homes heated and cooled by a combination of thermostats and microcomputer chips, watch pictures of world events transmitted by satellite unfold on their color television screen, and eat foods prepared and preserved by a wide array of technologies unknown to their parent's generation. When they become ill, they are treated with new pharmaceutical products that reflect twentieth-century advances in antibiotics, virology, or genetic engineering. For reasons of work, play, or family, millions of Americans routinely take commercial air transportation to destinations inside the country and in nations on the other side of the planet. The best-informed scientists that were alive at the beginning of the twentieth century would not recognize 10 percent of today's technologies.

Today's children—the next generation—will undoubtedly live in a significantly more scientific and technological culture. The rapid expansion of computer and robot technologies promises to relieve human beings of an ever larger share of routine and repetitious work. New advances in agriculture and plant genetics suggest that the perennial struggle to feed the population of the planet will require less and less effort. Advances in medicine, communications, and transportation may lead to significantly longer lives and to a world community able to talk and visit with one another. The curve of scientific and technological advance is still strongly positive.

The economic need for and value of a scientifically literate populace are well known. Science and technology have had a pervasive impact on both the methods of production and the products that are manufactured. The production of traditional industrial products like steel and the shaping of this and other metals into products has been largely automated. Work in the modern office is characterized by the machines and technologies utilized—word processors, data entry operators, data base managers, fax clerks, and photocopy technicians. The industrial challenges of the twenty-first century will be the manufacture of microcomputer chips, genetically engineered products, and new products yet to be invented. In this kind of economy, a basic understanding of science and technology will be the starting point for the development of the additional professional and technical skills needed to be competitive in an era of intense international economic competition.

Parallel to the need for a more scientifically literate work force, the economy of the twenty-first century will need a higher proportion of scientifically literate consumers. From the experience of the last two decades, it is clear that increased exposure to computers at work and school has stimulated a strong and growing home microprocessor market. As more products incorporate new technology, the information about the desirability, safety, and efficacy of those products will require a basic level of scientific literacy for comprehension. Some twentieth-century technologies like the irradiation of foods for preservation have never achieved a high level of commercial success due to public misunderstanding and resistance. A strong technology-based economy in the twenty-first century will require that a substantial portion of consumers be scientifically literate.

Looking ahead to the early decades of the twenty-first century, it is clear that national, state, and local political agendas will include an increasing number of important scientific and technological controversies. As new energy and biological technologies move toward the marketplace, there will be important public policy issues to be decided, and some of these issues may erupt into full-scale public controversies. The preservation of the democratic process demands that there be a sufficient number of citizens able to understand the issues, deliberate the alternatives, and adopt public policy.

The Scientific Literacy Construct

To understand the concept of scientific literacy, it is necessary to begin with a thorough understanding of the concept of "literacy" itself. The basic idea of literacy is to define a minimum level of reading and writing skills that an individual must have to participate in written communication. Literacy is most often presented as a dichotomy—literate versus illiterate—precisely because it is a threshold measure. The focus on a minimal knowledge level is inherent in the concept of literacy.

Historically, an individual was thought of as literate if he or she could read and write their own name. The person who had to sign his or her name with an "X" was defined as "illiterate." In recent decades, there has been a redefinition of basic literacy skills to include the ability to read a bus schedule, a loan agreement, or the instructions on a bottle of medicine. Adult educators often use the term "functional literacy" to refer to this new definition of the minimal skills needed to function in a contemporary industrial society. In the United States, for example, the social science and educational literature indicates that about a quarter of American are not "functionally literate."

The changing definition of literacy suggest some important characteristics of the basic concept. First, the level of skills needed to be considered literate changes over time. It is inherently a relative measure—not an absolute standard.

Second, given the diversity of social and economic systems on this planet, the same definition of functional literacy would not be appropriate for both advanced industrial societies and third-world agricultural societies. Any definition of literacy is inherently relative to the character of the society in which it is used.

Finally, the selection of a threshold level for the definition of literacy is not an exact science, but rather a judgment by those who understand a subject about the minimal acceptable level of knowledge or skill required to function in some set of roles in a specific society. In regard to basic literacy, for example, the literature indicates that there are several different tests or measures of functional literacy, reflecting the perceptions of each test author about the mix of skills necessary to function in society. A comparison of several of those tests, however, reveals that all are testing a common domain of skills and that there is a fair consensus on the kinds of skills and knowledge needed to be classified a functionally literate.

The problems associated with basic literacy in modern industrial societies are serious and relevant to our concerns about a broader public understanding of science and technology. For those millions of adults who are not functionally literate, the world of science is as distant as Pluto. And a very high proportion of the young people who are dropping out of school will join the ranks of the functionally illiterate.

In this context, science literacy should be viewed as the level of understanding of science and technology needed to function minimally as citizens and consumers in our society. A definition of scientific literacy does not imply an ideal, or even acceptable, level of understanding, but rather a minimal level. In previous studies (Miller, 1983*a*, 1987, 1989, 1992), I have argued that scientific literacy demands: (1) a basic vocabulary of scientific and technical terms and concepts, (2) an understanding of the process or methods of science for testing our models of reality, and (3) an understanding of the impact of science and technology on society. It may be useful to examine each of these dimensions in somewhat greater detail.

Understanding Basic Scientific Terms and Concepts

The first requirement for scientific literacy is an understanding of basic scientific and technical terms and constructs. If an individual cannot comprehend basic terms like atom, molecule, cell, gene, gravity, or radiation, then it would be nearly impossible for that person to follow much of the public discussion of scientific results or public policy issues pertaining to science and technology. In short, a minimal vocabulary of basic scientific constructs is necessary if one is to be scientifically literate.

As the use of standardized testing expanded during the 1950s and 1960s, especially in the United States, a number of tests were developed to measure a student's knowledge of basic scientific constructs (Buros, 1965). The majority of these tests have been used by teachers and school systems to evaluate individual students, to determine admission or placement, or for related academic counseling purposes. While some test-score summaries have been published by the Educational Testing Service (ETS) and other national U.S. testing services, these reports reflect only those students who plan to attend college or who have elected to take the test for some reason. Although very large numbers of tests are taken each year, the self-selected nature of the student populations involved continues to raise substantial problems for analysis and interpretation.

In the United States, the first and largest national data set that provided science achievement scores for a broad probability sample of American students was the National Assessment of Educational Progress (NAEP). For three decades, the periodic tests of the NAEP have collected measures of cognitive science knowledge from national samples of nine-, twelve-, and seventeen-year-olds. On the basis of five assessments between 1969 and 1986, NAEP found declining science achievement scores for all age groups and for almost all social and demographic groups (NAEP, 1988).

In contrast to the substantial effort to measure student knowledge of science and mathematics, there have been relatively few attempts to collect measures of scientific and technical knowledge from national samples of adults. In the United States, a 1957 study sponsored by the National Association of Science Writers (NASW) included a few knowledge items. In the early 1970s, the National Assessment conducted two studies of the science knowledge of young Americans aged twenty-six to thirty-five. The reasons for this dearth of adult knowledge measures are apparent. Most adults do not like to be tested on any subject matter, and the low levels of actual knowledge about science appears to heighten adult reluctance to be tested or measured in this regard.

Beginning in 1979, a series of surveys[1] of national probability samples of American adults has included a series of knowledge items. The number and scope of scientific constructs covered has increased in the continuing U.S. *Science and Engineering Indicators* series (National Science Board, 1993). Beginning with a joint U.S.–British study in 1988, a growing number of major industrial nations have sponsored national studies of the public understanding of science. The European Union has sponsored communitywide studies in 1989 and 1992. A national Canadian study was conducted in 1989. Japan conducted a national study in 1991 and is planing for a second cycle in early 1995. China conducted a national study in 1992 and has just completed the fieldwork for its 1994 national study. New Zealand conducted a national study in 1991, and Spain and Bulgaria conducted national studies in 1992. All of these studies included some items designed to measure the level of adult understanding of science, and there is a significant overlap in the selection of knowledge measures.

In the 1979 U.S. study (Miller, Prewitt, and Pearson, 1980), Prewitt and Miller introduced the idea of a series of two-part knowledge items that first asked each respondent to rate their own comprehension of a term. If the respondent reported that he or she had a clear understanding of the general sense about a term, they were asked to explain what the word or phrase meant in an open-ended response. If a respondent indicated that he or she had little understanding of a word, term, or concept, they were not asked for further explanation. This approach makes the collection of knowledge data from adult populations more acceptable to respondents, reduces unproductive time in the interview, and reduces the coding burden from open-ended answers.

In all of the U.S. surveys, an open-ended question about the meaning of scientific study has initiated the series. It is a difficult question for many respondents. Initially, there is a tendency for respondents to overestimate their level of knowledge about an item, but the effect of the first probe appears to have the effect of reducing the tendency to overestimate one's level of information. Using this observation, the scientific knowledge measure used in the 1979 and 1985 U.S. studies were based on the self-reported knowledge items for radiation, DNA, and GNP. These same items were repeated in the 1988 U.S. study, allowing a time series comparison. In both 1985 and 1988, 34 percent of American adults reported at least one clear understanding and one general sense on these three items, thus meeting a minimal criterion for scientific vocabulary.

In the 1988, 1990, and 1992 U.S. studies, an expanded number of scientific and technical knowledge items were included. Some of these items were presented in a true-false quiz format while others were posed as free-standing questions. While each study included fifteen to twenty scientific and technical knowledge items, nine of the items were identical to items used in parallel studies in Canada and the European Union and displayed sufficient variance to be useful in index construction. Respondents received one point each for a "true" response to the following items:

1. The oxygen we breathe comes from plants. Is that true or false?
2. Electrons are smaller than atoms. Is that true or false?

3. The continents on which we live have been moving their location for millions of years and will continue to move in the future. Is that true or false?
4. Human beings, as we know them today, developed form earlier species of animals. Is that true or false?
5. The universe began with a huge explosion. Is that true or false?

Respondents received one point for each "false" response to the following items:

1. Lasers work by focusing sound waves. Is that true or false?
2. The earliest human beings lived at the same time as the dinosaurs. Is that true or false?

In addition, respondents received one point for indicating that light travels faster than sound, and one point for a pair of answers indicating that the Earth goes around the Sun once a year.

In 1992, 40 percent of American adults met this standard (see Table 17.1). This level of vocabulary and concept comprehension means that three of five adults in the United States could not read and comprehend a newspaper or magazine story about a current scientific or technological controversy.

Understanding the Process of Science

The second requirement for scientific literacy is an understanding of the process of science, or the nature of the scientific approach. The critical question is whether a citizen knows enough about the process of scientific investigation to be able to distinguish between science and pseudoscience. While the number of controversies of this kind has been relatively small in the United States, it is

TABLE 17.1 Understanding Scientific Terms and Concepts, 1992, N = 2,001

The oxygen we breathe comes from plants. (True)	86
The continents on which we live have been moving their location for millions of years and will continue to move in the future. (True)	79
Which travels faster: Light or sound? (Light)	75
Knows that the Earth goes around the Sun once each year.	46
Electrons are smaller than atoms. (True)	46
The earliest human beings lived at the same time as the dinosaurs. (False)	45
Human beings, as we know them today, developed from earlier species of animals. (True)	45
The universe began with a huge explosion. (True)	38
Lasers work by focusing sound waves. (False)	37
Percentage answering six or more items correctly	40

important that a scientifically literate citizen be able to recognize when a study or report is based on science and when it represents other ways of thinking or knowing.

The systematic study of the public understanding of scientific thinking emerged in the 1930s as a result of Dewey's article "The Supreme Intellectual Obligation," in which he declared that the responsibility of science cannot be fulfilled by methods that are chiefly concerned with self-perpetuation of specialized science to the neglect of influencing the much larger number to adopt into the very make-up of their minds those attitudes of open-mindedness, intellectual integrity, observation, and interest in testing their opinions, that are characteristic of the scientific attitude (Dewey, 1934). Reflecting Dewey's basic charge, I. C. Davis, a prominent science educator of the period, defined an individual who has a scientific attitude as one who will: (1) show a willingness to change his or her opinion on the basis of new evidence; (2) search for the whole truth without prejudice; (3) have a concept of cause and effect relationships; and (4) make a habit of basing judgment on fact (Davis, 1935). Hoff (1936) and Noll (1935) offered similar definitions and began the task of developing items for use in testing. Virtually all of the empirical work before World War II focused on the development of the scientific attitude, or an understanding of the process of science.

Although numerous attempts have been made over several decades to define and measure scientific thinking among school-age and young adult populations, the first study to attempt to measure adult comprehension of the scientific process in the United States was a 1957 study[2] sponsored by the NASW. In that study, each of the 2,000 respondents were asked to define the meaning of scientific study, and the open-ended response was coded into a set of categories that reflected various levels of comprehension of the process of theory formulation and testing. Withey (1959) concluded that only about 12 percent of American adults could be said to have a reasonable understanding of the concept of scientific study.

The same open-ended question about the meaning of scientific study was repeated in national U.S. studies in 1979, 1985, 1988, 1990, and 1992. This item is an essential component in the measurement of the public understanding of the process of science. While the coded responses have shown a generally consistent pattern over their several uses, some of the responses are sufficiently vague or marginal to make reliance on this single item undesirable.[3] As a check against this open-ended question, each of the *Science and Engineering Indicators* since 1979 has included a set of items about astrology. One of the items asks respondents whether astrology is very scientific, moderately scientific, or not at all scientific.[4] To be classified as having a minimally acceptable level of understanding of the process of science, a respondent must be able to provide a satisfactory open-ended explanation of what it means to study something scientifically and to recognize astrology as not at all scientific.

Using this measure in the 1992 study, 14 percent of American adults displayed a minimally acceptable level of undertstanding of the process of science (see Table 17.2). Undoubtedly, a somewhat larger proportion think that they understand the process of science and some individual respondents may be able to mention some words that are associated with doing science, but when asked to explain the concept in an open-ended format, most cannot describe the process in terms that would be helpful to another adult or to a child in understanding the idea of a scientific study. To the extent that this measure includes error, and virtually all measures include some error, it is likely that this measure tends to slightly overestimate the real level of understanding.

TABLE 17.2 Understanding of the Process of Science, 1992, N = 2,001

Percent reporting:	
a clear understanding of scientific study	31%
a general sense of scientific study	50%
little understanding of scientific study	18%
Percent able to give acceptable open-ended definition of the meaning of scientific study	21%
Percent indicating that astrology is not at all scientific	62%
Percentage qualifying as having a minimal level of understanding of the process of science	14%

Understanding the Impact of Science on Society

The third requirement for scientific literacy is an understanding of the impact of science and technology on society broadly and on the daily life of individuals as consumers, parents, and citizens. As implemented in the 1979, 1985, 1988, 1990, and 1992 *Science and Engineering Indicators* (National Science Board, 1993) studies in the United States, this dimension has been measured with a series of items that might be thought of as "technological literacy" in a broader sense. While the items in the vocabulary dimension of scientific knowledge are important in understanding a wide range of scientific and technical material, some knowledge about computers or radioactivity might assist an individual in coping with some of the technologies and public policy issues that he or she may confront currently.

For the 1992 U.S. study and for comparative analysis with studies conducted in other countries, a three-part index was constructed to measure a social impact of technology dimension. Respondents received one point for recognizing that antibiotics do not kill viruses, one point for a correct understanding of the probability meaning of one-in-four, and one point for indicating that all radioactivity is not man-made. A fourth point was awarded to those respondents who indicated that they had a clear understanding[5] of the term "computer software."

Those citizens who were able to score three or more points on this simple index were classified as having a minimal understanding of the impact of science and technology on society. Using this very simple index, 31 percent of American adults met the minimal standard (see Table 17.3).

TABLE 17.3 Understanding of the Impact of Science and Technology on Society, 1992, N = 2,001

Percent correct response to:	
Is all radioactivity man-made or does some radioactivity occur naturally? (some occurs naturally)	73%
Probability interpretation of one-in-four	55%
Antibiotics kill viruses as well as bacteria. (False)	35%
Reports a clear understanding of the term "computer software"	25%
Percentage qualifying as having a minimal level of understanding of the impact of science and technology on society	31%

An Index of Scientific Literacy

To be scientifically literate, it is necessary to have a minimal vocabulary of basic scientific terms and concepts, a minimal understanding of the nature of scientific inquiry, and a minimal understanding of the impact of science on society. Combining the indicators of these three separate dimensions, approximately 7 percent of American adults qualified as scientifically literate in the 1992 study (see Table 17.4).

TABLE 17.4 Scientific Literacy in the United States, 1992, N = 2,001

Estimate of scientific literacy	6.5%
Understanding of scientific process	14.0%
Understanding of scientific terms and concepts	40.0%
Understanding of the impact of science and technology	31.2%

It is important to emphasize the threshold nature of this measure. The basic argument is that those individuals falling below this standard for scientific literacy lack some of the elements necessary for a citizen to understand the discussion and debate likely to surround a public policy controversy involving a scientific or technological issue. The requirements for classification as scientifically literate are minimal, and it is possible that some of the individuals who have managed to meet this standard might have a difficult time handling a relatively sophisticated scientific controversy, especially a dispute like the laetrile controversy in the United States a few years ago that revolved, in part, around what constituted adequate scientific testing or evidence of the medical efficacy of a substance.

The 1992 estimate of the level of scientific literacy in the United States shows no significant change from the results of previous studies in 1979, 1985, 1988, and 1990. In each of those years, national studies similar to the 1992 U.S. study were conducted and the estimated level of scientific literacy varied from 5 to 7 percent over the period from 1979 to 1992 (see Figure 17.1). Given the likely range of measurement error around these estimates of scientific literacy, it is safe to say that the level of scientific literacy among American adults ranged between 5 and 9 percent throughout the last decade.

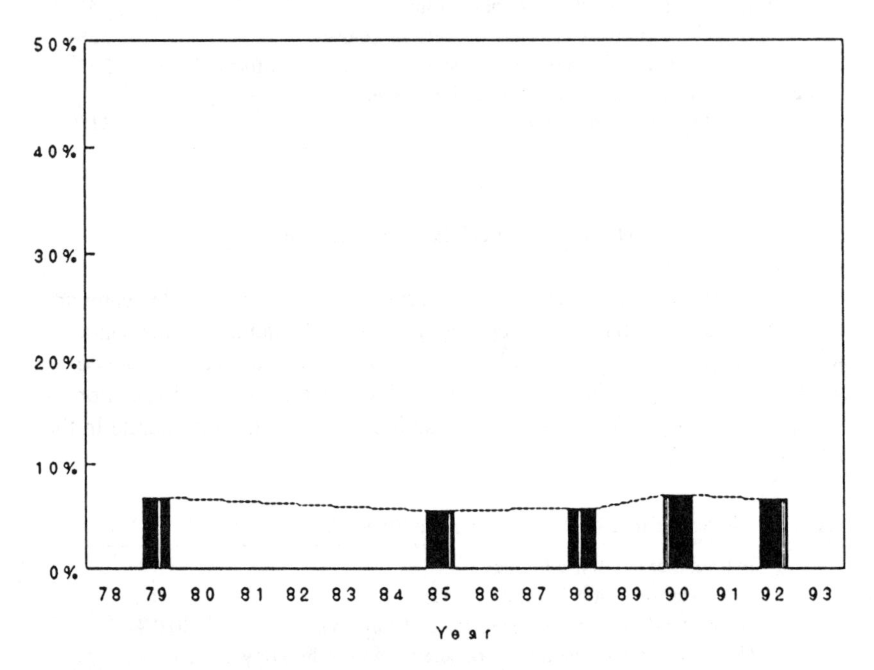

FIGURE 17.1 Percent of American Adults Qualifying as Scientifically Literate, 1979–92

The Distribution of Scientific Literacy

Having defined scientific literacy and estimated the percentage of American adults meeting that standard for 1992 and previous years, it is appropriate to turn to the issue of the distribution of scientific literacy in the United States.

Which segments of the public have the highest levels of scientific literacy? What is the impact of formal education, age, gender, and related factors on the distribution of scientific literacy?

In several previous studies by Miller and others, the level of formal education has been strongly and positively associated with scientific literacy and this pattern continued in the 1992 study. Fewer than 1 percent of Americans who did not complete high school met the criteria for scientific literacy, in contrast to 27 percent of Americans who have earned a graduate or professional degree (see Table 17.5). One in five Americans who earned a baccalaureate (but not a graduate degree) qualified as scientifically literate, but only 3 percent of adults with a high school diploma and no further degrees met that standard. Using the ordinal correlation coefficient gamma, the level of formal education completed accounts for 81 percent of the variance in the distribution of scientific literacy. This is similar to the relationships found in previous *Science and Engineering Indicators* (National Science Board, 1993) studies.

Prior to the 1990 U.S. study, respondents were asked about their exposure to college-level science courses, but no information was collected about the exposure of American adults to science and mathematics instruction during their high school years. Since most American adults have never enrolled in college and only about half of current high school graduates enter a postsecondary program, it is important to explore the impact of high school science and mathematics courses on the scientific literacy of adults.

Beginning with the 1990 U.S. study each respondent was asked whether he or she took a biology course, a chemistry course, or a physics course while in high school.[6] In 1992, a quarter of the respondents reported that they had never taken biology, chemistry, or physics in high school and 28 percent reported that they had taken only a biology course. Only 18 percent of American adults reported having taken all three courses—biology, chemistry, and physics—while in high school. A simple index of high school science course exposure was created, and it is strongly and positively related to scientific literacy (see Table 17.5). While less than 1 percent of American adults who had taken none of the three courses qualified as scientifically literate, 15 percent of adults who had taken all three high school science courses qualified. Using the ordinal correlation coefficient gamma, the number of high school science courses taken accounted for 63 percent of the variance in scientific literacy.

To measure exposure to high school mathematics instruction, each respondent in the 1992 U.S. study was asked to name the highest level of math taken during their high school program. Given the hierarchical nature of mathematics instruction in high schools, it is possible to infer the courses necessary to reach any given level. For this purpose, five levels were defined: first-year algebra, geometry, second-year algebra, precalculus, and calculus. While there may be some variation in the naming of these courses in some high schools, all

TABLE 17.5 Scientific Literacy by Educational Background, 1992

Background	*Percent Literate*	*N*
Highest Level of Formal Education		
Grade 9 or less	0.0	196
Grade 10 or 11	0.0	203
High school graduate	3.2	1,206
Baccalaureate	20.3	235
Graduate degree	27.1	161
Number of Years of High School Science Courses		
No high school science courses	0.5	492
One science course	2.9	633
Two science courses	11.0	522
Three or more science courses	14.7	354
Number of Years of High School Mathematics Courses		
No high school mathematics courses	0.8	567
One mathematics course	2.4	321
Two mathematics courses	6.4	311
Three mathematics courses	6.7	331
Four mathematics courses	14.9	335
Five mathematics courses	19.1	135
Number of College Science Courses		
No college-level science courses	1.3	1,351
One or two college-level science courses	10.5	252
Three or more college-level science courses	21.7	398
Science-Math Education Index		
Low exposure (four or fewer courses)	1.1	1,175
Medium exposure (five or eight courses)	6.7	467
High exposure (nine or more courses)	23.8	358

Note: The Science-Math Education Index is based on the sum of the number of high school science and mathematics courses taken and the number of college science courses in biology, chemistry, and physics reported. The Index excludes high school and college courses in science disciplines other than biology, chemistry, and physics.

national studies of mathematics course taking have found this to be the model structure of high school mathematics in the United States. Using this classification system 28 percent of American adults report no mathematics instruction at the algebra level, although many of these respondents reported participation in general math or business math courses. Only 7 percent of American adults have studied calculus during their high school experiences. A simple index of exposure to high school mathematics instruction was constructed and this index was strongly and positively associated with scientific literacy. While only 2 percent of adults with one year of algebra in high school qualified as scientifically literate, 19 percent of American adults who took high school calculus met the standard for scientific literacy. Using the ordinal correlation coefficient gamma, the number of high school math courses completed accounted for 60 percent of the variance in scientific literacy.

As in previous *Science and Engineering Indicators* studies, the number of college-level science courses taken was strongly and positively associated with scientific literacy in the United States (National Science Board, 1993). Each respondent who had completed high school was asked if he or she had taken any college-level courses in biology, chemistry, or physics and, if so, how many of these courses they had taken.[7] Approximately 70 percent of American adults have never taken a college-level science course, and only 1 percent of these adults qualified as scientifically literate (see Table 17.5). In contrast, 22 percent of adults who have taken three or more college-level science courses were scientifically literate. The number of college-level science courses taken accounted for 80 percent of the variance in scientific literacy.

Looking back over the relationships between the level of formal education, the number of high school science courses, the number of high school math courses, and the number of college-level science courses, it is clear that all four of these measures are strongly and positively associated with the distribution of scientific literacy. To provide a summary measure of each respondents's exposure to formal training in science and mathematics, an Index of Science and Mathematics Education was constructed. The number of high school science and math courses and the number of college science courses (up to maximum of ten) were summed, producing an Index that ranges from 0 to 18. For this analysis, the Index was trichotomized into three levels. Adults with four or fewer science and math courses were classified as having a "low exposure," reflecting a minimal high school program. Fifty-nine percent of American adults are included in this low exposure group. Adults with five to eight courses were classified as having a "medium exposure," which is the equivalent of a good high school program. About 23 percent of American adults are included in the medium exposure group. Adults with nine or more courses are likely to have had a good high school program and some college courses. Only 18 percent of American adults are included in this "high exposure" group.

The Index of Science and Mathematics Education is an excellent predictor of scientific literacy. Only 1 percent of American adults with a low exposure to science and mathematics education qualified as scientifically literate, while 24 percent of those with a high level of exposure were scientifically literate (see Table 17.5). This simple trichotomy accounted for 81 percent of the variance in the distribution of scientific literacy.

In summary, the preceding analysis demonstrates the general influence of formal education and the direct influence of science and mathematics instruction on the development of scientific literacy among adults in the United States. There can be little doubt that the primary engine driving the level of scientific literacy in the United States is the exposure of citizens to science and mathematics instruction, and these data indicate that vast numbers of American adults have had minimal formal training. Nearly 60 percent of American adults in 1992 qualified as having a "low exposure" to science and mathematics education.

Granted the primary role played by formal training in science and mathematics, it is useful to ask what other factors contribute to the distribution of scientific literacy. One factor is age, and it is often noted that many older Americans were educated prior to the last several decades of scientific achievement and advancement. The data from the 1992 U.S. study allow an exploration of the relative impact of age within the context of educational background. The results indicate that age has virtually no impact on the level of scientific literacy within levels of formal education (see Table 17.6). For example, about 5 percent of American adults who earned a high school diploma and no further degrees qualified as scientifically literate, except for those respondents aged forty-five and over. Among adults with a baccalaureate or graduate degree, there is no consistent association between age and scientific literacy. While a higher proportion of recent generations of Americans have been able to attend college and graduate or professional schools than older cohorts, there is no direct or significant influence of age itself on the distribution of scientific literacy.

In contrast to previous *Science and Engineering Indicators* studies, the gender of the respondent does not have a significant effect on the distribution of scientific literacy in the United States (National Science Board, 1993). In the 1992 U.S. study, 7 percent of adult men qualified as scientifically literate in contrast to 6 percent of adult women (see Table 17.7).

SCIENTIFIC LITERACY AND DEMOCRACY

Returning to the issue of scientific literacy in a democratic political system, the results from the 1992 U.S. study indicate that only 7 percent of American adults met a minimal test of scientific literacy. While 40 percent of American adults have a minimal vocabulary of scientific terms and concepts, only 14 percent can demonstrate a minimal understanding of the nature of scientific inquiry. This pattern suggests that a significant portion of American adults may be able to read news reports about scientific issues and controversies, but that they would be unable to recognize the difference between arguments based

TABLE 17.6 Scientific Literacy by Age and Educational Background, 1992

Education	*Age*	*Percent Literate*	*N*
Grade 11 or less	18–24	0.0	61
	24–34	0.0	48
	35–44	0.0	52
	45–64	0.0	104
	65+	0.0	133
High school graduate	18–24	4.9	196
	24–34	3.4	298
	35–44	5.3	260
	45–64	1.4	298
	65+	0.4	154
Baccalaureate	18–24	*	19
	24–34	19.7	76
	35–44	20.7	67
	45–64	19.3	49
	65+	9.7	23
Graduate degree	18–24	*	1
	24–34	31.0	37
	35–44	31.9	51
	45–64	25.7	56
	65+	*	16

* The percentage is not reported for cells with fewer than twenty cases.

on scientific investigation and arguments based on pseudoscience. For example, the pervasive public confusion over whether reports of unidentified flying objects indicate visitors from other civilizations in the universe reflects an inability to distinguish between conclusions based on scientific work and science fiction. If democratic processes are to survive into and through the twenty-first century, it is imperative that a significantly larger proportion of Americans and other citizens of industrial societies become and remain scientifically literate.

If we are to sustain our democratic political system in the decades ahead, it is essential that the proportion of scientifically literate citizens be increased. While a full discussion of the strategies to increase scientific literacy is beyond the scope of this chapter, it may be useful to outline some short-term and longer-term approaches to the problem of pervasive scientific illiteracy.

First, as demonstrated in the preceding analysis, scientific literacy is closely related to persistence in the formal study of science and mathematics. It is difficult to foster meaningful scientific literacy in adults who do not understand basic scientific concepts or the process of scientific inquiry. Beginning in

TABLE 17.7 Scientific Literacy by Gender and Educational Background, 1992

	Percent SL		*N*	
	Male	*Female*	*Male*	*Female*
All Adults	7.2	5.8	950	1,051
Highest Level of Formal Education				
Grade 9 or less	0.0	0.0	101	95
Grade 10 or 11	0.0	0.0	89	114
High school graduate	3.0	3.3	548	658
Baccalaureate	22.4	18.1	121	114
Graduate degree	27.7	26.3	91	71
Number of Years of High School Science Courses				
No high school science courses	0.6	0.4	224	269
One science course	2.7	3.0	265	368
Two science courses	10.8	11.1	222	299
Three or more science courses	15.2	13.8	239	115
Number of Years of High School Mathematics Courses				
No high school mathematics courses	1.0	0.6	257	310
One mathematics course	1.6	3.0	136	185
Two mathematics courses	3.6	8.5	136	175
Three mathematics courses	6.5	6.9	147	184
Four mathematics courses	16.6	12.9	183	152
Five mathematics courses	21.3	14.7	90	45
Number of College Science Courses				
No college-level science courses	0.9	1.5	608	743
One or two college-level science courses	11.2	9.8	128	123
Three or more college-level science courses	22.9	20.3	213	185
Science-Math Education Index				
Low exposure (four or fewer courses)	0.7	1.5	509	666
Medium exposure (five or eight courses)	6.1	7.3	234	233
High exposure (nine or more courses)	24.8	22.5	206	152

Note: The Science-Math Education Index is based on the sum of the number of high school science and mathematics courses taken and the number of college science courses in biology, chemistry, and physics reported. The Index excludes high school and college courses in science disciplines other than biology, chemistry, and physics.

the early 1960s, the number of science and mathematics courses required for high school graduation declined steadily for more than two decades. While there is some evidence of a modest increase in high school science and mathematics requirements, far too many American students have only a minimal exposure to formal science instruction. For most of the last two decades, only 20 percent of high school graduates have completed a year of physics and only half have completed a year of chemistry. American high school graduates have less formal schooling in science and mathematics than secondary school graduates in any other major industrial nation. The first step in increasing scientific literacy must be an increase in the minimum requirements for high school graduation to include three years of laboratory science and three years of mathematics, beginning with algebra.

Second, the provision of STS courses at the secondary level can provide an important linkage between the study of science and mathematics and the development of citizenship skills. It is important that STS courses do not replace basic science and mathematics courses in students' high school programs, but STS courses can utilize social science methods to help students understand the formulation of science and technology policy in increasingly specialized political systems. As I have argued elsewhere (Miller, 1983*b*), it is unlikely that science and technology policy issues will become electoral issues. Rather, science and technology issues will continue to be decided through legislative and administrative processes, or as Rosenau (1974) has labeled it, "citizenship between elections." It is essential that STS courses that address the formulation of public policy recognize this process and prepare students to be effective within it.

Third, nearly half of American students enter some form of postsecondary study. While college-level science courses have been an important source of scientific literacy among baccalaureate graduates in the United States, there is little evidence that this literacy is linked to an understanding of citizenship roles in complex specialized political systems. Presently, increased attention is being focused on the quality of the undergraduate science experience. In this review, it is important to focus some attention of the need for linkages between basic science courses and citizenship roles for both science and engineering majors and nonmajors. At the college level, smaller liberal arts colleges have often created interdisciplinary courses for this purpose, while larger universities have sometimes modified an existing political science or sociology courses to address this need. The construction of viable linkage courses at the college level should be an important item on the STS agenda.

Fourth, there are millions of adults who have finished their formal schooling, who have a high level of interest in science and technology policy makers, but who currently do not understand a number of important scientific and technological concepts. Many of these individuals have had some formal training in science and mathematics, although it may have been a few years ago, and appear to want to learn more about the science and technology that underlie contemporary issues. For example, attendance data from museum exhibits indicate that there are millions of adults seeking to better understand basic scientific and technical concepts and the issues to which they relate. Although these programs are usually classified as continuing or informal science education, there is the same need here for a linkage between the development of scientific literacy and the understanding of the formulation of public policy in the American political system. Ideally, some of what we learn from the development and implementation of programs at the high school and college levels can be usefully applied to the improvement of adult education programs.

National leaders in the United States and numerous other countries have recognized the importance of higher levels of scientific literacy, usually for economic purposes. There is a pervasive receptivity to efforts to improve the level of scientific literacy for secondary school graduates and the broader electorate of democratic societies. While it has been and will continue to be important to monitor the levels of student and adult understanding of science and technology in our respective societies, it is imperative that we utilize these results and our insights into the processes underlying these results to suggest approaches to increasing the proportion of our fellow citizens who are able to understand and participate in the formulation of public policy for our societies, including those issues involving significant scientific and technological components. Strong STS courses, linked to basic science and mathematics courses, can make a major contribution toward this objective.

NOTES

1. The 1979, 1981, 1985, 1988, 1990, and 1992 studies were sponsored by the NSF Science Indicators Unit. The 1983 study was sponsored by the Annenberg School of Communication at the University of Pennsylvania, with funding from the NSF Public Understanding of Science Program.

2. The fieldwork for this study was completed less than a month prior to the Soviet launch of *Sputnik I* in 1957, and thus represents the last measure of public understanding of science prior to the beginning of the Space Age. For a full description of the study and results see Davis (1958).

3. In the 1992 U.S. study, this open-ended question was coded independently by five graduate students. The intercoder reliability ranged from 0.78 to 0.82 (kappa).

4. In the 1988 British study, each respondent was asked to rate each of a series of items on a scale ranging from five for "very scientific" to one for "not at all scientific." Even though the five-point British scale differs slightly from the three-point U.S. scale, respondents in both studies had one clearly labeled choice of "not at all scientific" and respondents who understood the approach and processes of science should have been able to select the correct response in either question.

5. In the 1988 U.S. study, a follow-up probe asked the respondent to demonstrate their knowledge by providing an open-ended definition, and subsequent analysis of those results indicated that there was a very high correspondence between the self-assessment and the actual level of respondent knowledge about computer software, thus the probe was not employed in the 1990 or 1992 studies.

6. Questions that ask respondents to recall events that occurred several years ago often result in significant levels of measurement error, since many people cannot recall accurately past events. In this case, however, fewer than 2 percent of the respondents indicated that they could not recall whether they had taken any of these courses and, in debriefings, interviewers indicated that virtually all of the respondents appeared to feel certain about this information.

7. As noted above, recall questions are especially susceptible to error and become more susceptible as the period of recall becomes longer. In this case, those respondents who had taken a lot of college-level science courses often had difficulty recalling the precise number and would respond that they had taken "fifteen or twenty courses" as part of some specific program. Respondents who had taken to college-level science courses or only one or two appeared to be more certain about their response and interviewers reported that they could often name the course. For this analysis, the number of college-level science courses were grouped into those with no exposure to a college-level science course (68%), those with one or two college-level science courses (12%), and those with three or more college-level science courses (20%).

References

Almond, G. A. (1950). *The American people and foreign policy.* New York: Harcourt, Brace.

Barber, B. (1962). *Science and the social order.* New York: Free Press.

Buros, O. K. (1965). *The sixth mental measurements yearbook.* Highland Park, N.J.: Gryphon Press.

Davis, I. C. (1935). The measurement of scientific attitudes. *Science Education, 19,* 117–122.

Davis, R. C. (1958). *The public impact of science in the mass media.* Ann Arbor: University of Michigan Survey Research Center, Monograph No. 25.

Dewey, J. (1934). The supreme intellectual obligation. *Science Education, 18,* 1–4.

Hennessy, B. C. (1972). A headnote on the existence and study of public attitudes. In D. D. Nimmo and C. M. Bonjean (eds.), *Political attitudes and opinion change* (pp. 27–40). New York: McKay.

Hoff, A. G. (1936). A test for scientific attitude. *School Science and Mathematics, 36,* 763–770.

Miller, J. D. (1983*a*). Scientific literacy: A conceptual and empirical review. *Daedalus, 112*(2), 29–48.

Miller, J. D. (1983*b*). *The American people and science policy*. New York: Pergamon Press.

Miller, J. D. (1987). Scientific literacy in the United States. In D. Evered and M. O'Connor (eds.), *Communicating science to the public* (pp. 19–40). London: Wiley.

Miller, J. D. (1989, April). *Scientific literacy.* Paper presented at the meeting of the American Association for the Advancement of Science, San Francisco, California.

Miller, J. D. (1992). *The public understanding of science and technology in the United States, 1990.* Report to the National Science Foundation. Washington, D.C.: National Science Foundation, Division of Science Resources Studies.

Miller, J. D., Prewitt, K., and Pearson, R. (1980). *The attitudes of the U.S. public toward science and technology.* A report submitted to the National Science Foundation under NSF Grant # 8105662. DeKalb, Ill.: Public Opinion Laboratory.

National Assessment of Educational Progress. (1988). *The science report card: Elements of risk and recovery.* Princeton: Educational Testing Service.

National Science Board. (1993). *Science and engineering indicators—1993.* Washington, D.C.: U.S. Government Printing Office.

Noll, V. H. (1935). Measuring the scientific Attitude. *Journal of Abnormal and Social Psychology, 30,* 145–154.

Resenau, J. (1974). *Citizenship between elections.* New York: Free Press.

Withey, S. B. (1959). Public opinion about science and the scientist. *Public Opinion Quarterly, 23,* 382–388.

Chapter 18

STS: A Crossroads for Science Teacher Preparation and Development

Herbert K. Brunkhorst
David M. Andrews

Changes in Preparing New Science Teachers

The purpose of this chapter is to propose that an STS focus can provide the rationale for a science teacher preparation program that meets the needs of all science teacher candidates and addresses the recommendations being made by the National Academy of Sciences in their draft of the *National Standards for Science Education and Assessment* (National Committee on Science Education Standards and Assessment, 1992, 1993*a*, 1993*b*, 1994). The *Standards* are divided into several sections, two of which are "The Standards for Teaching" and "The Standards for Professional Development of Teachers." The "Standards for Teaching" focus on what teachers do. "The Standards for the Professional Development of Teachers" focus on how teachers develop professional knowledge and skill. Collectively, the standards help the science education community and the public to understand what science is essential and useful to an informed citizenry and the need for all students to leave school literate in science.

These recommendations will be discussed within the context of constructivism as a pedagogical paradigm for science teacher education followed by a proposed linkage between constructivism and STS for the purposes of restructuring schools and the requisite science teacher preparation. Some suggested program possibilities will be proposed to begin to construct a template for science teacher preparation and development appropriate for STS reform.

Almost a decade ago The Carnegie Report, *A Nation Prepared: Teachers for the Twenty-first Century* (1986), outlined those characteristics that will be necessary for students of the twenty-first century. These characteristics were: "coming to the workplace knowing how to figure out what they need to know, where to get it, and how to make meaning of it; having a feel for mathematical concepts and being able to apply these concepts to problem solving situations; having a cultivated creativity that leads to new problems, products and services before the competition; having the ability to work with other people in complex organizational environments; having the ability to think for themselves; acting independently and with others; and rendering critical judgments and being able to contribute in a constructive manner. Obviously if these were qualities that were recommended for students in the twenty-first century, these same qualities need to be considered as part of any teacher preparation program. Teachers will need to prepare students for an unexpected, non-routine world of the future. Schooling will have to refocus on learning rather than teaching, and from passive acquisition of facts to the active application of ideas to solve problems. This transition will make the role of the teacher more important, not less" (Carnegie Forum on Education and the Economy, 1986, p. 25).

There seems to be ample evidence that Americans love to learn. Americans do believe in education as evidenced by the scores of individuals who take on substantial learning projects, both formal and informal, to acquire the skills and knowledge they need in their lives. However, they choose to define it for themselves. What they do not seem to like is being taught or used, especially in ways that make them recall some of the unpleasant and unproductive experiences in their formal education. Many students do not expect to use what they learn in school. Classroom instruction, as currently practiced in many schools around the country, does not develop an aptitude for learning. Many have called for new ways of thinking about education. This new way of thinking demands and necessitates that new and current teachers have the kinds of experiences to make these dramatic shifts.

The National Science Education Standards are being designed to provide a coherent vision of what it means to be scientifically literate. The standards will describe what *all* students must understand and be able to do as a result of their cumulative learning experiences in science. Ensuring that science is learned by *all* students and that these students have some depth of understanding will require that science teachers have the requisite knowledge and skills that such an approach to science teaching will require. Such outcomes will require that practicing teachers and those preparing to be the science teachers of tomorrow will themselves undergo some major shifts in thinking. In a recent Los Angeles *Times* editorial Bruce Alberts, president of the National Academy of Sciences, points out that we do not need large numbers of scientists, but rather large numbers of scientifically trained citizens.

Champagne and Klopfer (1984) have offered a comprehensive review that sheds light on the science literacy question. Central to the research is the fact that large numbers of students "learn" school science but without real understanding or "belief" in what they purport to know. There are major breaks between school science and the real world. Completing more complex/abstract science courses does not help students, even college students, to develop real understanding of the world.

"Some day, Americans will follow science events in the news with the same interest and excitement as they follow business or sports" (Hazen, 1993, p. 15). This statement described a vision that Robert Hazan spoke of in a speech titled, *Not for Scientists Only: Redefining the Crisis in Science Education.* He challenges the conventional view of his colleagues in the sciences who see the science education crisis as not having enough young people interested in going into the sciences and not getting enough rigorous science. Their solution is to make science more rigorous at an earlier and earlier age. Unfortunately in their attempts at rigor they have produced a kind of "rigor mortis" for the overwhelming majority of people who feel they can never comprehend science. Hazen advocates a broader approach that helps the remaining 99 percent of students. He recommends that a broader approach would help most students understand a wide range of issues that affect their lives. Students rarely see the relevance or purpose to the science they are asked to learn. Hazen also recommends tailoring teaching in such a way that no vocabulary would be used that is not in the newspaper, contains no complicated mathematics, and places a constant emphasis on science as part of everyday life (p. 17). He recommends that we precede rigor by letting everyone understand that science is relevant and interesting. Hazen uses the analogy, "You don't dig a deep well until you survey the land to see where to put the well" (p. 19).

CONSTRUCTIVISM AND SCIENCE TEACHER EDUCATION

Currently most educational scholars espouse the idea that knowledge is constructed. Individuals interpret what they experience in terms of what they already know and believe. This is frequently the root of the problem with large lecture classes, especially in science. The instructor inevitably has an experience base far beyond that of the students in the class. What each student does with the information being presented is oftentimes quite different from what the instructor's perceptions are about the presented information.

As indicated in earlier chapters, constructivism is not a theory of teaching, rather it is a theory of learning. It does not prescribe how to teach. Rather, constructivism provides a framework for how we should search for evidence concerning what it is we have taught. How well the teacher teaches becomes inseparable from how well the student learns. Constructivism provides us with a constant reminder that students construct their own realities with the information they gather and whether we like it or not they do so with little regard for our "truths." It is critical that teachers who accept the constructivist paradigm become sensitive to the wide range of ways that students can interpret a lesson.

A constructivist approach to science teaching will involve new roles and demands on teachers of science and on science teacher educators. A broader set of roles, behaviors, and strategies must be mastered. The old paradigm views knowledge as being constructed by experts and transmitted to students who in turn might become teachers. Just as students think for themselves, so do teachers. We do not yet have a template of constructivist teacher education. However, we as science teacher educators can model constructivist teaching. It would necessarily involve participation in a community of learners. It is through such social interaction that discussions, debates, and arguments can play a critical role in challenging the adequacy of old concepts and beliefs about the teaching and learning process. It is in this way that we as science teacher educators, or science teacher candidates, can better understand our own teaching.

A university science program should help students figure out what they know, what they need to know, where to get it, and how to make meaning of it. The content in these courses should build on the existing concepts of the students within the various science disciplines and should include reflection on their own understanding of these concepts. If we develop these types of students, then the same characteristics can be reframed to describe the types of teachers needed to support this type of learning. A constructivist paradigm has tremendous implications for teacher preparation just as does the concept of science for all students.

STS: The Crossroads

STS represents an essential theme of science education for all (Aikenhead, 1980; Bybee, 1986; Harms and Yager, 1981; Hurd, 1986). Brunkhorst and Yager (1986, 1990) propose that an STS focus can provide the kind of redefinition that is needed to address the criteria expressed in the National Science Standards. Roy (1983) has termed the science, technology, society focus as the "glue" of science for most people.

In a paradigm shift toward constructivism and STS, the teacher's role is away from directing all classroom discourse to guiding student activity. The challenge is whether or not we can establish an atmosphere where we engage in risk-taking, wondering, conjecturing, testing, and arguing. Since STS issues are complex, an STS focus can provide a somewhat level playing field where a variety of expertise can be brought to the discussion by both student and teacher. STS topics and the likelihood that no one knows all the answers, begins to move teacher candidates away from "show it to me" and "tell me how to do it" kinds of statements. Students begin to see knowledge as not necessarily fixed and objective, but rather, as evolving and moving toward knowledge that is somewhat subjective, interpretive, and contextualized. We begin to see education moving from an organization based on teaching to an organization based on learning or from the transmission of knowledge to the generation of knowledge.

The combination of an STS school curriculum with a constructivist instructional methodology may prove to be the engine of a successful revolution in education (McFadden, 1991). Several science educators have recommended the application of constructivism to teaching (Cleminson, 1990; Cheung and Taylor, 1991; Vosniadou, 1991). If one accepts a constructivist view of learning, then one can assume that science teacher candidates will bring preconceptions about science teaching and learning into science methods classrooms (Hewson, Zeicher, Tabachnick, Blomker, and Toolin, 1992). It suggests the necessity of placing the teacher-to-be in a central position. A successful preparation program provides an environment where the preservice teacher identifies the problems and has the experiences of dealing with them. Stofflett (1994) reports that teacher candidates construct science content knowledge and conceptual change pedagogy in an interrelated manner (p. 788).

The importance of fidelity to a teaching approach and its faithful demonstration to teacher candidates cannot be underestimated. We as science teacher educators will have to also "walk the talk." Research has demonstrated that teacher candidates cannot be expected to coherently represent a pedagogy they have never personally experienced. Yet they are frequently asked to do so (Stoddart, Stofflett, and Gomez, 1992). This could be said to apply to both constructivist pedagogy and an STS focus toward science teaching. Teacher candidates hold behavioristic epistemologies (Ball, 1989; Hollingsworth, 1989). Therefore, they are likely to see conceptual change theory as implausible. These candidates need to see a plausible approach to helping students construct knowledge, and to the use of STS as a focus for addressing science curriculum issues.

Research on the assessment of STS beliefs and views of teachers may be relevant to the design of science teacher preparation programs. If understanding is critical, then teachers need to understand their knowledge base involving STS issues. Zoeller et al. (1990) suggests there be a special effort to enhance the STS literacy domain within teacher preparation and development programs. University science courses are taught without considering what entails understanding, how to cultivate it, and how to gather systematic evidence that you have it. Most scientist-teachers think that conceptualizing material for students results in the students having an exact replica of the knowledge originally in the head of the instructor. Gess-Newsome and Lederman (1993) suggest restructuring undergraduate science content courses in terms of the content taught and the pedagogy used since the subject matter structures of teacher candidates are more highly developed into didactic schema. Lederman (1986) states the sine qua non of instruction prescribes that teachers can teach only what they understand. Hence, science teachers need an adequate understanding of STS interactions in order to help students develop such understandings (Rubba, 1989).

Currently an analogous situation is occurring in the medical education community. Many medical educators are examining and questioning the fact-absorbing method of teaching so prevalent in medical schools today. Are there any analogies to what we see in science education today? Perhaps we can help each other. Marlys H. Witte, professor of surgery at the University of Arizona Medical School suggests that medical school textbooks would contain many blank pages that would remind medical students and teachers of all that their profession has yet to learn (Blum, 1993).

Dr. Witte says, "Books may spend eight to ten pages on pancreatic cancer without ever telling the reader that we just do not know that much about it—what causes it, how to stop it, how to treat it without tremendous discomfort" (Blum, 1993, p. A21). She continues, "This is a sham. In deference to our ignorance, there should be at least four or five pages left blank to be filled in by future doctors. How do our science methods texts look? Do our books have any blank pages? Do our science books have any blank pages? Interestingly, our student's laboratory books are empty until we have them copy verse and chapter from an existing text. We need to acknowledge that science education like medicine is rife with ignorance, failure, and chaos" (Blum, 1993, p. A24).

Dr. Witte runs a popular program called the Curriculum on Medical Ignorance. Students, professors, visiting professors, and local high school teachers and students participate in activities designed to teach them, in Dr. Witte's words to "define and even celebrate what they know they don't know, don't know they don't know, think they know, but don't, and also don't know that they know" (Blum, 1993, p. A24). She has wondering rounds and pondering sessions. Perhaps we as scientists and science educators need to consider borrowing "a page" from our colleagues in the medical education profession.

The National Research Council continues to define standards for science teaching which, in turn, provide a solid direction for science teacher preparation. Much of what the Council has proposed is in line with the STS approach to teaching science. The professional development of teachers should include opportunities to address issues, events, problems, or significant topics in science that are of interest and are relevant to the learner. Teacher preparation, in a constructivist manner, must build on the teacher candidate's existing science knowledge, skills, and attitudes. The NRC further makes some recommendations as to the kinds of science content that teachers need to be prepared to teach. The science and technology component of recommended content establishes "useful connections between the natural world and the designed world and offers essential decision making abilities." Educating future science teachers to effectively teach science in a changing society driven largely by technological advances has certain implications for the science teacher preparation program. Certainly, the undergraduate science program must reflect the importance of science in a context that makes the key connections to societal and technological issues; it is the science education that students receive in their undergraduate courses that heavily influences how they perceive science and its relation to the world.

If we are successful in producing teachers who are capable of making connections between scientific concepts and societal and technological issues, as many of the science education reforms are urging, we must alter the way we prepare science teachers and specifically the way in which science concepts are presented in the undergraduate science classroom. Science teacher preparation programs must include science content courses based on a constructivist teaching-learning model, use alternative assessment strategies, present science from an interdisciplinary viewpoint, and make relevant connections to everyday life.

The Commission on Teacher Credentialing in the state of California has taken a major step in this direction by requiring all credentialed teachers in science to not only have a science concentration (biology, chemistry, geosciences, and physics) but have a year of laboratory-based science in the other three content areas. Such an effort begins to support the kind of breadth and depth required of science teachers to address the current national science education reforms.

A New Type of Science Teacher Preparation

There are some specific problems that have been identified with the current way in which most science teacher candidates are prepared. These problems include: (1) mimicking a single cooperating teacher; (2) having experience with only one discipline; (3) being perceived by students as not being a "real" teacher; (4) conflicts between the philosophy of the school-building faculty and university faculty; (5) difficulties in finding school-building faculty that are familiar with reform and restructuring issues; and (6) difficulties in getting university faculty to recognize restructuring issues involving curriculum and teaching improvement. The current paradigm continues to focus on "experts" dispensing knowledge and answers being more valued than the asking of good scientific questions. When was the last time a scientist seriously reflected on his or her knowledge before transmitting it to students in lecture?

As we stated earlier, though a template for a constructivist teacher preparation program has yet to be developed, there are general characteristics or qualities that one might expect to see. These include: (1) teacher as learner; (2) an STS focus that places both teacher and student into the position of learner, especially when integrating into another's field; (3) the establishment of a community of learners; (4) developing one's own strategies for sense-making requiring reflection on what one knows and how they know it: (5) syllabus construction involving the community of learners; (6) active learning that involves becoming engaged in the process of bringing new knowledge and new ways of knowing to bear on a widening range of increasingly more difficult problems; and (7) a variety of assessments so that if a pattern is revealed, we can have some confidence in the validity of the results.

Teachers find that the actual observing of demonstration lessons is more beneficial than videotapes in understanding pedagogical decisions and connections made by the teacher. Practicing teachers visiting California SS&C classrooms to see how coordinated/integrated science is actually taught have noted the value of such firsthand observations. Such demonstration classrooms have become a rich resource for several science teacher preparation programs around the state. Such sites can help science teacher candidates understand the integration of a constructivist approach using STS issues as the context in obtaining science literacy for all students.

Abell (1992) reports a strategy for helping science teacher candidates make sense of the numerous articles that they read and discuss during the course of a semester. It involves a reading reaction sheet that has the student examine their ideas before reading begins, relating the focus of the article on what they already know. Assessment needs to be adjusted as well. The pursuit of meaning through reflection is the ultimate goal of science teacher preparation. Such strategies can provide ways to help science teaching candidates construct meaning about science teaching and learning. Anyone who gives students reading assignments must be concerned that students read the material, make sense of it, and can apply it to classroom activities and discussions. The instructional questions become: How can we find out what sense students are making of the assigned readings? and How can we help them in the process of constructing meaning from textual material? Research has shown that activities that involve writing and reading have been more effective at increasing learning, thinking skills, and inquisitive attitudes, than activities involving reading alone (Langer and Applebee, 1987). In addition, student journals or learning logs have proven to be an effective aid to student learning (Mayher, Lester, and Pradl, 1983).

Discrepancies exist between how science teacher candidates and science teacher educators perceive "order and control" (Latz, 1992). This has tremendous implications for how well we address the needs of our students as they prepare to enter the science classroom as first-year teachers. We need to increase our research efforts to identify the characteristics that distinguish a master teacher from the novice in terms of management. Some of the major distinguishing characteristics seem to be related to time, confidence, and support (Loughran, 1994). If science classrooms are going to be investigative arenas where students are actively engaged, we as science teacher educators need to consider the additional management responsibilities in such an environment. Acquiring and organizing materials, physically arranging them, determining smooth transitions into and out of activities are all management issues to consider in the STS classroom. Prior research suggests teaching experience is a key factor in the development of management skills (Hollingsworth, 1989; Pigge, 1978). Once again modeling, demonstration classrooms, can play key roles. As currently practiced, during the practicum experience the actual impact of demands still rests with the cooperating teacher. This calls for innovative ways to get teacher candidates into more real-life situations.

The Golden State Examinations in 1993 provided laboratory performance tasks in biology with an evolutionary problem focusing on ecological and STS issues. The exam overwhelmed most students with the requirement to think (Filson, 1994). Some conclusions that have implications for science teacher preparation are: (1) if we want to know how students actually perceive scientific concepts, we need to give them more experiences solving open-ended problems; (2) if we want to improve instruction and thereby understanding, we need to identify common misconceptions and give these ideas more attention; and (3) if students are to think more critically, we need to provide experiences and practice in analyzing and contributing to the experimental design.

Preservice teachers in science and other fields frequently complete their programs without ever confronting their beliefs about science teaching and science content. Recognizing and challenging one's beliefs is an important aspect of learning and teaching and needs to be addressed in science teacher preparation programs if one is to have any influence on teacher candidate views and beliefs. The teaching portfolio can help candidates reflect on their understandings of what it means to be a science teacher, what it means to teach science for all, what it means to teach science for meaning, what it means to make science relevant to the lives of students. We need to help our students confront the gap between expectations and realities. Many teachers, regardless of their previous experience, are surprised at how the dynamics of discipline and management change when students are involved and interested in their own learning. How to prepare teachers to make that leap after being subjected to the typical models of didactic teaching is a major challenge for the science teacher educator.

The realities of our scientific and technological age requires redoubling efforts to promote global awareness and a broadly defined literacy among teachers and students. Breaking out of the mold will require either a change in the way university science is taught or a reduction in the degree to which prospective teachers take science at the university. To acquire understanding in science or science pedagogical content knowledge one needs to ask the questions: What do I know? How do I know it? How well do I know it? and, Is it true? All teachers at all levels will have to open the world of science to themselves. Arnold Arons (1983) has presented the challenge: "How can we expect the students to cease running from anything that makes them feel intellectually insecure if we persist in running ourselves?" (p. 110). We need to build a culture of science teaching that requires careful reflection on the teaching and learning process. We need to be certain that we do not separate the knowing from the finding out.

A useful model for what a science teacher preparation program might look like in the context of STS and constructivism appears in the SAGE (Science and Global Education) Group Model suggested by Haury, Conwell, Fuller, Lydecker, and Staley (1987). This model has as its primary theme fostering the development of the intellectual skills necessary for understanding issues and resolving problems with an emphasis on the personal search for understanding and individual efficacy. The model's implications for science teacher preparation involve all teacher candidates in efforts to: (1) be well-grounded in the liberal arts including world history from both a Western and non-Western tradition; (2) be familiar with a historical overview of science and technology including the background for analysis and consideration of both present day and future issues and problems; and (3) be experienced with a study of world systems (economic, political, technological and ecological) based on organizing patterns including their relationship to human welfare. The model recommends that science courses be integrated providing greater breadth so that science can be better understood from a global perspective. It recommends that science be shown as a human enterprise having a history and value structure; and as an enterprise that strongly influences human interaction and well being while generating knowledge. Students should be exposed to the nature of scientific inquiry and science as a social force. And, students should examine the interactions between science and technology.

Potential coursework might include: global perspectives seminar, environmental studies, world health and nutrition, philosophy and ethics in science, an integrated science and social studies methods course, cultural diversity, decision-making, and critical analysis. The model suggests the secondary science methods class be restructured and titled, The Role of Science Education in the Schools. It is suggested that professional studies be integrated into field experiences whenever possible and that partnerships between schools and universities be established to set up communities of learners. These ideas have been built on previously developed and published guidelines for science teacher preparation (Ritz and Mechling, 1984).

Science teachers of "tomorrow" will have to integrate learning, teaching, and assessment as never before. Teachers themselves must engage in ongoing assessment of their teaching and student learning. Those in all aspects of science teacher preparation need to ask, "Why do I want students to know this? Will knowing this information make a difference in their lives? Is it important? Do the students understand this? How can I be sure? How frequently does our agenda as professors dictate our student's ability to think?" There will be little gained if we reinvent science teacher preparation without regard for the nature and ethos of science as it is practiced today. We need to recognize that science information becomes knowledge only when we demonstrate an ability to use it. As Paul DeHart Hurd (1994), has so eloquently stated, "The goal is to live a life of choice not chance" (p. 2).

REFERENCES

Abell, S. K. (1992). Helping science methods students construct meaning from text. *Journal of Science Teacher Education*, *3*(1), 11–15.

Aikenhead, G. S. (1980). *Science in social issues: Implications for teaching*. Ottawa, Canada: Science Council of Canada.

American Association for the Advancement of Science. (1993). *Benchmarks for science literacy: Part I: Achieving science literacy: Project 2061*. Washington, D.C.: Author.

Arons, A. (1983). Achieving wider scientific literacy. *Daedalus*, *112*, 91–122.

Ball, D. L. (1989). *Breaking with experience in learning to teach mathematics: The role of the preservice methods course*. (Issue Paper 89-10). East Lansing, Mich.: Michigan State University, NCRTE.

Blum, D. E. (1993, January 27). Arizona professor offers curriculum on ignorance to give medical education a wake-up call. *Chronicle of Higher Education*, pp. A21, A24.

Brunkhorst, H. K., and Yager, R. E. (1986). A new rationale for science education–1985. *School Science and Mathematics*, *86*(5), 364–374.

Brunkhorst, H. K., and Yager, R. E. (1990). Beneficiaries or victims. *School Science and Mathematics*, *90*(1), 61–69.

Bybee, R. W. (1986). Science-Technology-Society: An essential theme for science education. In R. K. James (ed.), *1985 AETS yearbook—Science, technology and society: Resources for science educators* (pp. 3–14). Columbus, Ohio: SMEAC Information Reference Center and Association for the Education of Teachers in Science.

Carnegie Forum on Education and the Economy. (1986). *A nation prepared: Teachers for the twenty-first century. A report of the Task Force on Teaching as a Profession*. Washington, D.C.: Author.

Champagne, A. B., and Klopfer, L. E. (1984). Research in science education: The cognitive psychology perspective. In D. Holdzkom and P. B. Lutz (eds.), *Research within reach: Science education* (pp. 171–189). Charleston, W.V.: Research and Development Interpretation Service, Appalachia Educational Laboratory.

Cheung, K. C., and Taylor, R. (1991). Towards a humanistic constructivist model of science learning: Changing perspectives and research implications. *Journal of Curriculum Studies*, *23*(1), 21–40.

Cleminson, A. (1990). Establishing an epistemological base for science teaching in the light of contemporary notions of the nature of science and of how children learn science. *Journal of Research in Science Teaching*, *27*(5), 429–446.

Filson, R. (1994). Student misconceptions revealed by performance testing. *California Science Teachers Journal*, 2–8.

Gess-Newsome, J., and Lederman, N. G. (1993). Preservice biology teachers knowledge structures as a function of professional teacher education: A year-long assessment. *Science Education*, *77*, 25–45.

Harms, N. C., and Yager, R. E. (eds.). (1981). *What research says to the science teacher, Volume 3*. Washington, D.C.: National Science Teachers Association.

Haury, D. L., Conwell, C., Fuller, M. W., Lydecker, A. M., and Staley, F. A. (1987). Preparing science teachers with a global perspective: Recommendations of the science and global education (SAGE) group. *Access*, 5–10.

Hazen, R. M. (1994, September). Address given by Robert Hazen, at the proceedings of Howard Hughes Medical Institute's Precollege Program Directors Meeting, Science Museums: Creating Partnerships in Science Education, Bethesda, Maryland.

Hewson, P. W., Zeicher, K. M., Tabachnick, B. R., Blomker, K. B., and Toolin, R. E. (1992, April). *A conceptual change approach to science teacher education at the University of Wisconsin-Madison*. Paper presented at the Annual Meeting of the American Educational Research Association, San Francisco.

Hollingsworth, S. (1989). Prior beliefs and cognitive change in learning to teach. *American Educational Research Journal*, *262*, 160–189.

Hurd, P. DeH. (1986). A rationale for a science, technology and society theme in science education. In R. Bybee (ed.), *Science, technology, and society* (pp. 94–104). Washington, D.C.: National Science Teachers Association.

Hurd, P. DeH. (1994). Toward a new vision of general education in the sciences. *California Science Teachers Journal*, 1–2.

Langer, J. A., and Applebee, A. N. (1987). *How writing shapes thinking: A study of teaching and learning*. Urbana, Ill.: National Council of Teachers of English.

Latz, M. (1992). Preservice teachers perceptions and concerns abut classroom management and discipline: A qualitative investigation. *Journal of Science Teacher Education*, *3*(1), 1–4.

Lederman, N. G. (1986). Relating teaching behavior and classroom climate to changes in students' conceptions of the nature of science. *Science Education*, *70*, 3–19.

Loughran, J. (1994). Bridging the gap: An analysis of the needs of second-year science teachers. *Science Education*, *78*(4), 365–386.

Mayher, J. S., Lester, N., and Pradl, G. M. (1983). *Learning to write: Writing to learn.* Portsmouth, NH: Boyton/Cook.

McFadden, C. P. (1991). Towards an STS school curriculum. *Science Education*, *75*(4), 457–469.

National Committee on Science Education Standards and Assessment. (1992). *National science education standards: A sampler.* Washington, D.C.: National Research Council.

National Committee on Science Education Standards and Assessment. (1993*a*). *National science education standards: An enhanced sampler.* Washington, D.C.: National Research Council.

National Committee on Science Education Standards and Assessment. (1993*b*). *National science education standards: July '93 progress report.* Washington, D.C.: National Research Council.

National Committee on Science Education Standards and Assessment. (1994). *National science education standards, discussion summary, September 12, 1994.* Washington, D.C.: National Research Council.

Pigge, F. L. (1978). Teacher competencies: Needs, proficiency, and where proficiency was developed. *Journal of Teacher Education*, *29*(4), 70–76.

Ritz, W. C., and Mechling, K. R. (1984). *Standards for the preparation and certification of teachers of science, K–12.* Washington, D.C.: National Science Teachers Association.

Roy, R. (1983, May 19). Math and science education: Glue not included. *The Christian Science Monitor*, pp. 36–37.

Rubba, P. (1989). *The effects of an STS teacher education unit on the STS content achievement and participation in actions on STS issues by preservice science teachers.* Paper presented at the 1989 Annual Meeting of the National Association for Research in Science Teaching, San Francisco.

Stoddart, T., Stofflett, R. T., and Gomez, M. L. (1994, April). *Breaking the didactic teaching-learning-teaching cycle: Reconstructing teacher knowledge.* Paper presented at the Annual Meeting of the American Educational Research Association, San Francisco.

Stofflett, R. T. (1994). The accommodation of science pedagogical knowledge and the application of conceptual change constructs to teacher education. *Journal of Research in Science Teaching*, *31*(8), 787–810.

Vosniadou, S. (1991). Designing curricula for conceptual restructuring: Lessons from the study of knowledge acquisition in astronomy. *Journal of Curriculum Studies*, *23*(3), 219–237.

Zoeller, U., Ebenezer, J., Morley, K., Paras, S., Sandberg, V., West, C., Walters, T., and Tans, H. (1990). Goals attainment in science-technology-society (STS) education and reality: The case of British Columbia. *Science Education*, *74*(1), 19–36.

CHAPTER 19

THE IDENTIFICATION OF SCIENCE CONCEPTS IN STS TEACHING THAT ARE REALLY ESSENTIAL

Martha Lutz

SCIENTIFIC LITERACY

Teaching science in schools represents a societal decision that science is critical for living in our society. In what sense is science critical? The current trend is to speak vaguely of "scientific literacy," and to assert that no person will be able to survive or make reasonable decisions in today's complex technological environment unless he or she has had some necessary but undefined dose of science education.

Leaders in science education generally agree that scientific literacy is desirable. Many assert that it is the goal of science education. Teachers and administrators accept the belief that scientific literacy, whatever it may be, is the current objective for modern science curricula (Champagne, 1989). David Hershey (1990) claims that scientific literacy *can* be achieved, and states that this will come about partly through linking science to everyday life and current events.

The swell of agreement and belief and desirability breaks apart when it is time to define scientific literacy, or to set criteria for achieving or evaluating scientific literacy.

CRITERIA FOR SCIENTIFIC LITERACY

What is scientific literacy? Without going into detail, it is sufficient to indicate here that the sum total of all the knowledge that all the specialists would consider necessary for a person to know in order to be considered scientifically literate would take a lifetime—several lifetimes—to acquire. Even within the past year (1992–1993) there have been two documents published that are teetering on the brink of being merely checklists for scientific literacy. Quite formidable checklists, too. Project 2061 has published a set of benchmarks that students should have reached at various grade levels. Scope, Sequence, and Coordination has published and is revising The Content Core, a set of topics that provide a framework for developing curricula for grades six through twelve.

Both these documents have some real value: both are the product of collaboration among specialists in science and science education. Both include concepts and principles and facts that would certainly be good for all graduating high school students to know. Both also include knowledge that, while noble and valid, is not required for a full, profitable, and well-informed life in our modern world. It would be hard to argue that a person who does not understand Faraday induction is unfit for life in this modern world. It would be possible for a practicing scientist, a person by definition "scientifically literate," to do excellent work in biology, be well informed on societal issues, take an active part in local recycling programs, and so on, and still never even have heard of Faraday induction.

Neither of the above-mentioned documents says anything about the importance of teaching children to wash their hands after using the toilet and before eating. Many adults do not understand the importance of this simple act. And yet it could be argued to be a cornerstone of scientific literacy. It touches on issues such as: diversity of life, microbiology, exponential growth, populations, microhabitats, health, evolution, and many others. Can a person be truly scientifically literate if he or she does not understand the issues underlying the importance of washing one's hands? Does mere exposure to such things as Planck's constant or the Doppler effect constitute scientific literacy?

Somehow, in our efforts to define and set criteria for this highly desirable state called scientific literacy, we have gone astray. We have been betrayed into believing that the definition must include a comprehensive list of essential facts, concepts, and principles, and that learning all of this science content is part of the criteria for being considered scientifically literate. Should the criteria for scientific literacy include a required list of essential content? To answer that question, we need to reexamine the basic concept of literacy.

THE CONCEPT OF LITERACY

Science education is not the only field struggling to define or redefine the concept of literacy. In his review championing two books on cultural literacy, Gagnon (1987) defends the books by overcoming the artificial schism between what he calls "content people" and "methods people." He emphasizes that the advocates of cultural literacy want to see history taught in context; he reminds the reader that any fact worth knowing will shed light on at least one important concept—a general concept that is illustrated by the specific fact. Rather than generating a list of information that one must know to be culturally literate, Gagnon defends the view that a small number of concrete experiences should be given intensive study.

It is clear that even if Gagnon is not familiar with science, he would appreciate Poincare's remark that "Science is constructed of facts, as a house is of stones. But a collection of facts is no more a science than a heap of stones is a house." Process without substance may be sterile, but substance without process is dead. But if content does not define scientific literacy, what does?

What is literacy? What is—if I may call it this—literary literacy?

Teachers of English literature could surely come up with just as daunting a checklist as the science educators, if they were asked to name all the great works of literature that are indispensable for a truly literate person. I would insist that all students read everything by E. M. Forster, and a few biographical works about him, as well. Someone else might say that no one can be considered literate who has not read Shakespeare. But is it necessary to read *all* of Shakespeare to qualify as having read Shakespeare? All the plays? All the sonnets? When is enough enough?

If all the opinions of all the experts were taken into consideration, even a severely abridged list of important literature would take a lifetime—or several lifetimes—to read through just once. Fortunately for our students, the definition of literary literacy stops short of being a consensus checklist of everything that everybody must read to be considered literate. It is considered sufficient to know how to read, and to demonstrate an ability to read, understand, analyze, and evaluate a somewhat random sampling of writing from the almost infinite list of essential literature. One *need not* read everything on the list, one need only *demonstrate a capability and a desire to be able to read and analyze* a piece of literature. That is the critical difference between the successful working definition of literary literacy and the unsuccessful attempts to create a working definition for scientific literacy.

That is where we went astray: with the perceived need to actually master all the *science* on the combined list of all the science experts. That would be analogous to reading all the literature on the combined list of all the literature experts. We need to follow their lead: scientific literacy is not defined by a checklist, no matter how comprehensive or valid it may be. Scientific literacy should instead be defined by *the ability to* acquire *scientific knowledge, and to comprehend, apply, and evaluate that knowledge.*

The phrase "scientific literacy" has become so caught up in semantic politics that it has almost become meaningless. What differentiates scientific literacy from any other kind of literacy? One criterion is that scientific literacy ought to imply *the ability to think in terms of evidence supporting a reasoned conclusion.* If you turn your head from side to side as you try to locate the source of a sound, you are practicing the mechanisms of a scientific approach. A child who watches crickets on an August evening, and then announces that the males sing and the females can hear the males, is practicing science and is therefore a scientifically literate person, in at least one sense.

SCIENTIFIC LITERACY—REVISITED

Science is a way of looking at and organizing phenomena and concepts. This definition is accessible to society. Scientists may be viewed as a subset of people who specialize in attempting to make sense of the natural world through a combination of logic, creativity, and experimental manipulation. In this sense, a medical doctor, a physicist, and a child raising a caterpillar are all, equally, scientists. And a toddler's mother, trying to find a way of potty training that "works," is also a scientist if she develops an idea, tries it, modifies it based on a preliminary set of results, and eventually achieves satisfactory results: a child who uses the toilet independently.

The complement to this viewpoint, anyone who understands the need to evaluate evidence before making a decision is scientifically literate, even if he or she is not officially a scientist. And a so-called scientist who refuses to examine evidence before arriving at a conclusion is no scientist—however, many paper PhDs he or she may possess that attest to content knowledge.

Therefore, scientific literacy must not be a list of concepts and phenomena—not even those agreed on by the acknowledged experts in each discipline. If anything at all should be listed, the processes of science belong on the list: observing, classifying, measuring, interpreting data, inferring, communicating, controlling variables, developing models and theories, hypothesizing, and predicting (The Content Core, 1992). A person who has learned to apply these processes has achieved scientific literacy in the same sense that a person who has learned to read, and to analyze and evaluate written material, is literate with respect to literature.

WHERE DO SPECIFIC CONCEPTS FIT IN?

If scientific literacy is to be redefined as learning science processes, rather than mastering a specific list of essential concepts, where and how do specific concepts fit in? Should all students be required to study any kind of list of truly essential concepts? If we really believe that scientific literacy is not dependent on the particular details of scientific knowledge acquired by the individual, then the answer to the above question is "no."

And yet, there will be an outcry from scientists—and nonscientists—who insist that there are some things that simply *must* be learned, if any science at all is to be learned. These people are correct: some things must be learned. For example, can we imagine teaching biology without going into specifics about evolution? Or chemistry without teaching about atomic structure? Physics—or any science discipline—without investigating energy? Once again, the answer is "no."

The two points raised above may appear mutually contradictory. They are not. A process-oriented approach to science teaching will *inevitably* lead to direct encounters with the major themes in the science disciplines. Trying to study any topic in biology without encountering the theory of evolution (and its mechanism: natural selection) would be like trying to travel from Iowa to Illinois without crossing the Mississippi River. The truly essential concepts do not need to be built into the curriculum because they are built into the discipline under study. No matter what particular topic the students and teachers tackle, the essential concepts will be encountered as surely as the Iowa-Illinois traveler encounters the Mississippi River.

Can we justify a curriculum in which one of the major determinants of what topic will be investigated is student interest? Even if teachers are empowered to guide and affect student choice, is this simply too radical a proposition? Not if we accept Reinsmith's (1993) "Ten Fundamental Truths about Learning." His third point is directly relevant to the question of how much input students should be allowed to have in determining their own curriculum. "*Students will learn only what they have some proclivity for or interest in.* Find out what a person likes, then help him build around it. Once interest exists learning is possible, and teaching kicks in. We waste enormous quantities of time (and money) giving students learning tasks for which they have no interest or readiness, boring them and frustrating ourselves in the process" (p. 8).

Does this imply that students should have free rein to determine what, if anything, they will study? Not at all; but it makes explicit something that should have been obvious for several decades: our curriculum has not yet been a success in engaging and retaining student interest in science. We have ignored the obvious: people will work hardest and invest most in what they have an interest in. This is human nature.

It is also possible (as the advertising industry has shown us dramatically since the advent of television) to create, or stimulate, interest. A primary factor in creating interest is providing a relevant context. The critical job of the teacher, then, is to facilitate the articulation of a relevant context to engage student interest in the study of a science concept. This choice of a relevant context is characteristic of STS teaching. It consists of defining a question or phenomenon that the students will investigate.

Essential concepts in science education are emergent properties of the dynamic interaction between teacher, students, and a well-chosen carefully defined topic for investigation. These emergent properties will be understood and retained better than if they were passively prescribed by a top-down curriculum. Studies have shown that students are more likely to comprehend and retain knowledge that they construct and that has personal relevance for them. Thus, we need to take into consideration the belief that it is correct to teach less content, but with better retention.

What is the use of teaching children great checklists of facts, concepts, rate laws, and so forth, when most students have forgotten their meaning within a few days and have forgotten they ever heard of them by the time they start their adult lives? Even practicing scientists do not carry as a mental repertoire all the essential science concepts listed in, for example, the Content Core.

Random content items listed in the Content Core were used to create a Scientific Literacy Survey. Presumably, items listed in the Content Core may serve as criteria for determining who is scientifically literate: all our high school graduates are expected to be familiar with all the items listed in that book. Individuals who fail to demonstrate at least some familiarity with the essential concepts listed in the Content Core are assumed to be scientifically illiterate.

The Scientific Literacy Survey was given to practicing scientists, members of the faculty of the University of Iowa. Preliminary data from the Scientific Literacy Survey include:

1. Nine out of fifteen biologists have never heard of or are only vaguely familiar with the concept of Faraday Induction (which dates back to the mid-1800s).
2. Six out of fifteen biologists do not know the difference between igneous and metamorphic rocks.
3. Ten out of fifteen biologists are unfamiliar or only vaguely familiar with the concept of biomes. Is this acceptable, even for people who consider themselves primarily molecular biologists? Or are these ten PhD research biologists actually scientifically illiterate?

If practicing scientists cannot meet the standards set for our high school graduates, then either the criteria for scientific literacy are invalid, or our practicing scientists are scientifically illiterate. Is something wrong with our scientists, or with our criteria for scientific literacy? The latter seems far more likely.

Practicing scientists achieve scientific literacy by meeting criteria of being able to manipulate science processes. Essential concepts from different disciplines may or may not be part of their repertoires; the ability to investigate, comprehend, analyze, and evaluate new information are always part of a scientist's repertoire.

The repertoire of practicing scientists does not only include the processes of science. Also included are qualities of desire and playfulness. These cannot be generated by any formulae. Such qualities can only be nurtured where they already exist.

Textbook-driven and content-oriented teaching squelches desire and frowns on playfulness. Context-driven and issue-oriented teaching can nurture the fragile and necessary characteristics that promote a positive interaction with the study of science.

How do we go about determining what to teach in a science class, at any level? Knowledge and skills that are never used, and perhaps not well understood even at the time of teaching, are almost guaranteed to be forgotten. Knowledge and skills that are learned in a context and applied to real-life situations are likely to be understood and remembered. This implies that we need to teach science that has personal relevance to the students. We need to provide opportunities for students to construct their own understanding of the science concepts they are working with, and also opportunities for them to apply their newly constructed knowledge in real-life situations.

The above description is the working definition of an STS teaching strategy.

A Method for Achieving Scientific Literacy

Scientific literacy cannot be defined by what we know, but by how we go about finding out when we do not know. Is it possible to teach students "how to find out"? Teaching students how to find out is the goal of the teaching approach known as STS:

How Do You Find Out when You Do Not Know?

1. Teachers should help students define a specific question or phenomenon that they want to find out about.
2. The students should learn to brainstorm, and to create a list of resources that will help them gather evidence. If necessary, they should be prepared to design experiments to gather data (evidence).
3. The students should use the resources that they have identified, and perform the experiments that they have designed.
4. They should learn to sit down and think about their data: analyze, synthesize, and evaluate. Creativity should be encouraged at every step.
5. Last, and perhaps most significant, the students should take action.

The steps outlined above define STS teaching.

THE NEW SCIENTIFIC LITERACY

A student who can do all the above can be considered just as scientifically literate as the average scientist. And the teacher who uses the teaching strategy described above is practicing STS: a strategy in which the teacher acts as a facilitator while the students become actively involved in seeking information and applying it to real-life situations and issues.

It is not necessary to identify and list essential science concepts. The genuinely essential concepts relevant at any point in the maturation of a specified science discipline are inextricably entwined in any representative issue or phenomenon the students may choose to investigate.

Our anxiety to create lists of essentials to be "covered" reveals our human foible: an inability to have faith that things are what they are. Any concept so trivial that it is easily avoided in an STS classroom is probably not essential. Any concept so general and important that we would label it "essential" is something we could scarcely avoid even if we tried.

Traditional teaching, using lists of "essentials" in a manner of crutchlike dependency, is really just a symptom of our human infirmity. Let us have faith that $A = A$, that something essential really is essential, and that therefore it is as unavoidable as the Mississippi River. Let us try a teaching strategy that produces students who know *how to find out* and *how to examine and evaluate evidence.*

These students may or may not be familiar with Faraday induction, igneous rocks, or biomes; but rest assured that if they ever need to know about such things, they will have the tools to find out. And that is what scientific literacy is really all about: just ask the scientists.

REFERENCES

Champagne, A. B. (1989). Defining scientific literacy. *Educational Leadership, 47*(2), 85–86.

Gagnon, P. (1987). Content counts. *American Educator*, Winter, 40–46.

Hershey, D. R. (1990). Science literacy is possible. *BioScience, 40*(7), 482.

Reinsmith, W. A. (1993). Ten fundamental truths about learning. *The National Teaching and Learning Forum*, 2(4), 7–8.

National Science Teachers Association. (1992). *Scope, sequence, and coordination of secondary school science: Volume I, the content core. A guide for curriculum designers*. Washington, D.C.: Author.

CHAPTER 20

STS PROMOTES THE REJOINING OF TECHNOLOGY AND SCIENCE

Karen F. Zuga

HOW THE PROBLEM AROSE

Except for the few technology and science educators who advocate the STS instructional approach, the communities of technology and science educators have been passing in our schools and universities as two ships pass silently in the night without speaking to each other about their relationship. In fact, they have grown so apart due to their specializations, they have become unable to speak each other's language (Lux, 1984). They are from two different cultures (Snow, 1959). Enmeshed in their own cultures, technology and science teachers are perpetuating this artificial separation of technology and science in the next generation, the students whom they teach.

The once unified activities of technology and science grew apart as the modern era gave impetus to technology and science, causing each area of endeavor to develop different modes of operating. The activity of science evolved to become abstract and theoretical while the activity of technology became concrete and practical, separating theory from practice (Dewey, 1925; Whitehead, 1925; Snow, 1959). As each endeavor added to the knowledge base, the complexity of each forced a separate evolution by forcing those who were technologists and scientists to specialize in one or the other. Educators have merely mimicked the separation of technology and science by creating subject matter distinctions and proceeding to further the separation of technology and science by passing on the abstract/concrete language differences to new generations of students (Kowal, 1991).

This artificially and human created separation did not go unnoticed, however. Philosophers and educators such as Dewey (1925), Whitehead (1925), and Snow (1959) recognized the error of this path and often criticized, particularly from an educational point of view, the separation of technology and science.

They questioned, as we question today, the value of this separation for the education of all children. Today, we still are faced with the inability of teacher educators, teachers, and students to speak across the artificial boundaries of technology and science. This state of affairs leaves us all with an incomplete and less sophisticated understanding of interrelationships and functioning of technology and science. As predicted by the philosophers of the early part of the century, this situation we have created and perpetuated can handicap our ability to progress.

It is time to rejoin technology and science. Technology and science educators are beginning to realize that the reunification is occurring at the forefront of investigation as new areas of inquiry such as biotechnology are created by researchers. Technology and science educators' task in classrooms is to search for ways in which to open and create a discussion that helps to integrate technology and science for students.

The Roles of Technology and Science

Just to set the record straight some working definitions of technology and science are in order. These are, perhaps, oversimplifications, but they should be addressed—as a focus on the distinctions between technology and science, rather than merely citing definitive statements about what technology and science are. As the rejoining of technology and science is approached, it is good to note that technology is viewed as a human endeavor to modify one's environment and science as a human endeavor to explain one's environment (Lux, 1984). Each construct is a human activity, directed by humans to fulfill our needs and wants; therefore, we control technology and science. The distinction between the constructs is made as a simple modification versus as an explanation of human effort. It is not a dichotomy nor a dualism—for other possibly parallel efforts not to be discussed here that deal with aesthetics, philosophy, and other constructs, enter into human endeavors. The distinction is made merely as a means of distinguishing the role of each in a complex pattern of relationships.

Whether we choose to modify our environment lightly or to alter it radically in order to live within it, technology has been of such fundamental value to people that we have chosen to let it become transparent to ourselves (Ihde, 1990) until something about our technology needs to be fixed (Kranzberg, 1991). As long as the technological systems that maintain a comfortable existence in modern buildings work quietly and efficiently in the background, we do not bother to notice or to try to learn about them. Let them stop working or create a health hazard to humans because of the way in which they were designed to work, then we begin to take notice of the technology, but only briefly, until we resolve the problem and our chosen technology slips quietly into the background noise of contemporary society. The truth of the matter is that without our use of technology all of our life support systems that we have grown so accustomed to using would disappear. There would be no more easily accessible food, shelter, and clothing, no tools to create these things, and no language for record keeping. We would reduce ourselves to the state of animals in nature. As for our choice to develop and use science, we had the basic necessities of life before this codified body of knowledge and means of investigating and explaining came into formal existence (Dewey, 1925). Ancient humans were first technologists in order to meet basic needs (Kranzberg, 1991; Selby, 1993). After those needs were met, then the time to investigate nature and explain why the technology worked arose. Today, if we were to find ourselves in a catastrophe that had severed us from our use of modern technology, we would once again find ourselves in a situation that would demand that we all become technologists providing for the basic needs to sustain life.

An integral part of human existence has been to be technological. The need and urge to modify the environment in order to provide basic necessities and to provide additional comfort has been ever present in the history of humanity. People, both women and men, have been technologists from the time that they began to communicate; clothe themselves; seek shelter; and develop tools for hunting, gathering, and storing their own food supplies. The materials that people fashioned into tools and finished goods through technical processes were not only created to fulfill basic needs, but they were also created to provide for extra comfort and to be pleasing to the eye. Also integral to humans has been inquiry, the desire to know and to explain (Winner, 1977). In order to maintain and improve our use of technology people employed a form of inquiry providing for description and explanation (Kranzberg, 1991). It is from this inquiry that some cultures chose to develop forms of technology and science.

Around the world different cultures have made choices about how to develop and use technology, creating variability in technological development. "No evidence exists of any culture that has gotten away without some attempt to understand, alter, and exploit nature. No evidence can be found of any human society that has not employed tools and techniques. The interesting question is why the modern West has proceeded along these paths with virtually no sense of limits" (Winner, 1997, p. 134). The choices that people made were related to their worldviews and belief systems. Western culture grew out of the Greco-Roman tradition of utilizing technology to modify the environment.

More important, the Greek tradition provided the philosophical foundation for the separation of technology and science by the very creation of the idea of science (Dewey, 1925). While a separation of technology and science was not achieved by the Greeks, they provided the intellectual framework for the possibility. Aristotle's hierarchical separation of knowledge into the structure of *theoria*, *praxis*, and *poiesis* as theoretical, practical, and productive ways of knowing provided a framework for looking at different ways to construct knowledge (Hickman, 1990). This kind of structure acknowledged that there were different forms of knowledge with the potential for developing different structures of knowledge. About theory and the ensuing growth of science, Dewey wrote: "This accomplishment was beyond the reach of artist and artisan. For no matter how solid the content of their own observations and beliefs about natural events, that content was bound down to occasions of origin and use. The relations they recognized were of local areas in time and place. Subject-matter underwent a certain amount of distortion when it was lifted out of its context, and placed in a realm of eternal forms" (1925, p. 105). The ability to create knowledge, which was liberated from the practical and a result of thought and logic, set the stage for the creation of science and the eventual separation of technology and science for the purposes of formal study. This separation was necessary for the creation of science and the eventual development of Western culture and Western technology. The ability to make this intellectual separation provided the impetus for the development of Western civilization as we now know it.

What is unfortunate about the separation is the enduring practice of employing Greek hierarchies in dualisms and continuums (Lux, 1981). Many of the Greek dualisms and continuums referring to such things as gender, love, and thought have left us with an enduring hierarchy of value attached to the concepts (Fox Keller, 1985). In the case of technology and science, science was thought to have greater value than art or technology. Science is theoretical while the arts are practical and technology is mere production. The Greek hierarchy is an underlying factor in the way in which we have thought about the relationship of science and technology (Hickman, 1990).

In society and in the profession of education we live with the legacy of this hierarchy today, even though our practice with respect to technology and science blurs the distinctions between science and technology in professional literature, general knowledge of society, and specialized knowledge of research laboratories. Lux (1984) found many instances of references to science that imply science *and* technology in professional literature. In the popular media there are numerous examples of references to science that imply or are confused with technology. In research laboratories the granting of science PhDs and research projects for the development of new instruments of measurement is common. There are also new fields of study such as biotechnology which have been created with the intention of modifying our environment through concerted research in explanatory science. Just what separates an act of technology from science is not clear (Kranzberg, 1991).

The separation of technology and science for purposes of study and the tendency to blur the distinctions between technology and science by practice haunts our curriculum in schools. We have created the separate subject matters of technology education and science education that are on unequal footing. Moreover, we have created an unrealistic representation of the role of technology and science in our society and the relationships between technology and science. Our confusion can hardly be helpful for our students.

THE RELATIONSHIP BETWEEN TECHNOLOGY AND SCIENCE

There are many ways to try to describe the relationships between technology and science. Symbiotic, two sides of the same coin, science as the branch of inquiry for technology, and science as the impetus for advanced technological development are some of them that come to mind. The kind of relationship one expresses tends to conform to one's beliefs. The problem is to define the relationships in ways that enable people to begin to see the connections. Without the ability to see the connections and to articulate the relationships, meaningful unification of the abstract and the concrete will be difficult, if not impossible, to achieve. Judging by practice in schools and universities, this has not been achieved. Even those who advocate STS in the schools continue to bring up the cultural differences and question the ways in which we can bridge the gap (Gaskell, 1982; Bauer, 1990). In many schools and universities technologists and scientists are still passing one another in the foggy night with no sustained communication.

This artificial separation of technology and science for the purposes of analysis and study does not exist in application. The relationship between technology and science has always been symbiotic, both efforts are necessary for the advancement of knowledge in either human endeavor (Kranzberg, 1991). This aspect of the relationship is even more crucial in contemporary society.

Our use of technology without the investigation and explanation provided by scientific inquiry could (and has at some times and in some cultures) easily become a fixed set of rituals. When the knowledge base of technology becomes fixed, acts of modifying the environment could become sets of prescriptions that are handed down from generation to generation without question. This knowledge, then, is passed on as a series of ritualistic steps to follow with a few embellishments added as time passes. Examples of these types of technological prescriptions can be found in old recipes, particularly those that are passed on in family traditions. Loss of the process is another danger inherent in not understanding why the technology works. We have the physical evidence of many crafts and material processes such as metalworking in ancient Africa, which have been lost due to lost records or the death of an oral tradition. Historically, what has happened most frequently is that people have developed a means of making or doing something before they have understood why and how what they are doing works (DeVore, 1980s). The need or desire has created the impetus for processes and/or products that have been developed before the explanation of why and how the solution is working is known. The knowledge and skill base of technology often is self-sustaining (DeVore, 1980s; Lux, 1984), but it also provides the fertile ground for description and explanation in science. "That the sciences were born of the arts—the physical sciences of the crafts and technologies of healing, navigation, war and working of wood, metals, leather, flax and wool; the mental sciences of the arts of political management—is, I suppose, an admitted fact. The distinctively intellectual attitude which marks scientific inquiry was generated in efforts at controlling persons and things so that consequences, issues, outcomes would be more stable and assured. . . . In responding to things not in their immediate qualities but for the sake of ulterior results, immediate qualities are dimmed, while those features which are signs, indices of something else, are distinguished" (Dewey, 1925, pp. 107–108). From the way in which we have employed technology scientists have been able to identify abstract concepts that can be tested in other contexts for the generation of new knowledge. That information, in turn, is recaptured by technologists and added to the knowledge gained by technological innovation as they continue to develop new technology. In this way technologists and scientists continue to feed each other symbiotically.

Just as our scientific research can sustain technological development, our technological development often promotes and sustains advances in scientific research. The description and explanation offered by scientists depends on the further refinement of instrumentation as technological devices. From the efforts of Galileo applying the telescope to view the moon to the very complex web of technology that sustains contemporary research in science, instrumentation is necessary for scientists (Hickman, 1990). Instrumentation as technology opens up vistas that exceed the capability of the naked eye. Without instrumentation scientists are severely handicapped in their ability to identify, describe, explain, and test. Without contemporary instrumentation in the form of magnifying lenses, photography, computerization, and robotics, it is hard to imagine how we could conduct science today.

In the end, each effort is essential to the other. Contemporary efforts in technology and science are beginning to demonstrate clearly the ways in which the two efforts are entangled. Edison is often credited with establishing some of the first science labs that were dedicated to employing investigation for the purpose of product development. This is now the norm in industry. The recent development of aerogels is another case in point. Commercial laboratory scientists developed a material that is light in weight and strong. The task for them now is to create a use for the new material. Research in genetics is another example of the blurring distinction between scientist and technologist. The technological applications of new genetic information is well known and can drive the scientific effort. People who have an interest in modifying their environment by creating designer vegetables as well as eliminating genetic diseases are developing the information and capability to do that with both scientific investigation and technological design efforts (Bishop, 1993; Kevles, 1993).

Relationships between the activities of scientists and technologists are being established and utilized faster than educators can identify and describe them. The key concern for educators is how to best unify the teaching of technology and science so that current and future generations of students understand the relationships and can work with them. Technology and science educators have been aware of the need to bridge the divide between the two subject matter cultures. Science education literature has been filled with a call for the integration of the subjects of technology and science, but the execution of the integration has not been a widespread phenomenon (AAAS, 1989; Bybee, 1991; Hurd, 1991; Rubba, 1991).

Two Educational Cultures and Languages

We are not doing very well communicating across the cultural divide. There are some initial efforts, but often these come directly from one field or the other (Gilliom, Helgeson, and Zuga, 1991). There are a number of exemplary programs housed in science, social studies, and technology education. In most of these efforts, curricula have been created by one or more members of the same field (Bybee, 1991; Gaskell, 1982; Hurd, 1991; Splittgerber, 1991). Having both technology educators and science educators trying to work together points out the cultural differences. In a review of the problems of integrating disciplines and school subjects as the subjects of STS, May (1991) writes: "The meanings of subject matters are mediated and dispersed through the cultural metaphors, language, material artifacts, and strategies we use in their presentation in university and school classrooms, popular culture, mass media, and other social institutions. In sum, whether we speak of disciplines or school subjects, we are speaking of arbitrary, amorphous, and temporal human constructions" (p. 76). Science educators and technology educators deal with their respective fields in such different ways that their working together has not been widespread. Even projects which attempt to bridge the cultural divide suffer from the rigidity of each culture (Gaskell, 1982).

In 1990 a science education colleague and I teamed up with science, technology, and special education teachers from a local school in order to create a science and technology course for all of the students. It did not take us too long to realize that technology and science educators from the university and the school spoke about their knowledge base and the curriculum in very different ways. Interestingly, we all could identify topics of mutual interest and concern, but our interpretations of those topics were distinct. We fell into our academic cultural stereotypes and proceeded to act from those biases (Bauer, 1990; Rubba, 1991). The end result of the project was a program for the high school teacher responsible for the learning disabled classes. Perhaps, due to her not being a member of the two cultures, she was able to straddle the cultural divide and implement a technology and science program for her students. The science and technology teachers willingly helped her, but could not identify how they could conduct a combined program.

Let me use an example of the cultural divide which comes from the discussions held by the teacher educators. As we began to discuss curriculum there was an immediate issue. We did not approach curriculum in the same way. Technology educators focus heavily on creating curriculum for many reasons. They do not have many textbooks in use because school administrators do not buy as many texts for their laboratories; technology education textbooks are often outdated as they are being published because they are concerned with representing contemporary use of technology; and there is a firm commitment to laboratory and actual application as instruction (Zuga, 1991*b*). For their own survival and to carry on the mission of technology education teachers are asked to be curriculum specialists. Science educators appear to rely more on a curriculum outlined for them in their textbooks. They have a more stable curriculum structure (Hurd, 1991). They do not question this curriculum in the way that technology educators do. Our first discussions about content were rather amusing. I wanted to delve into the content and my colleague came from a culture which saw this as rather a strange and unnecessary behavior at best. Discussing this difference led immediately to another one which was later to be mirrored in the group discussions with teachers. Given an agreed-on topic such as optics, each of us proceeded with it in a different manner. The technology culture took over in me as I listed processes and devices as key content and the science culture took over in my colleague as she listed principles. We were in a concrete/abstract communication gap. Later, as the teachers continued to identify topics and content for an integrated program, the same concrete/abstract concept gap kept appearing. Ultimately, those teachers who came from the two cultures opted out of implementation of the course. They cited time commitments, but I suspect that they could not fully identify with what was created as a combined course. Even my colleague and I let go of the project as the demands of our own cultures moved us on to other efforts. The cultural divide is deep and wide in academe.

Given the cultural divide, the question of the value of having the two groups try to hammer out curriculum is raised. Why not continue to have each field of study identify integrated technology and science curriculum for themselves? According to Hurd (1991) the call for integrating technology into science education has been present since the inception of science education. He documents a number of failed attempts at getting widespread adoption of technology in science education. In addition, Bybee (1991) and Rubba (1991) document the research that deals with teachers' conceptions and acceptance of integrated technology and science. They also paint a dismal picture of the ability to institute and sustain an integration of technology and science through science education. In the related subjects of social studies and technology education Splittgerber (1991) and Zuga (1991*a*) discuss only technology and science themes within social studies and no widespread or sustained integration efforts within technology education. In social studies education Heath (1992) states that technology and science units are often the result of teachers without support working alone. Instances of integrated technology and science in technology education are reported as similar isolated occurrences by Brusic (1992).

By far, the science education community has had the greatest experience with integrating technology and science both in its implementation and research about the success of implementation. Yet, integration of technology in science education appears to be failing. Based on his research, Rubba (1991) offers a constructivist explanation about the failure of integrating technology and science. Essentially, his thesis is that teachers with no background and experience with technology cannot be expected to implement an integrated program. His thesis is not in contradiction with the thesis I have offered here with respect to the cultural divide. It is very difficult for either technology or science educators to overcome their own culture in order to devise and implement integrated programs. Examples of technology concepts held by science educators cited by Bybee (1991) appear to be no more than a list of social implications for science. Splittgerber's (1991) technology and science themes in social studies are just that, themes about technology and science that impinge on the accepted content of social studies. As a technology educator, I know of no widespread emphasis on integrating science in technology education. Perhaps, it is lack of knowledge and perspective of the other culture that is causing the failure. Any single curriculum innovator or team of innovators from one field do not know enough about both fields in order to do justice to representing both technology and science in their respective subject matter.

Currently, the Project 2061 team is attempting to integrate both mathematics and technology into their science education curriculum plans (American Association for the Advancement of Science, 1989). While they are not developing an STS curriculum per se, they have been busy trying to incorporate subject matter that has not been a tradition in science education. Their goals are to be commended. They have, however, limited the participation of those outside of the science community to one of critique as the curriculum standards are being identified. Perhaps, they, too, recognize the difficulty associated with the cultural divide and they did not have the time to try to bridge the communication gap that separates the two cultures.

The problem that curriculum created by a majority of members of one field will face, however, is that it will remain a comprehensive field specific curriculum with a smattering of concepts from other subjects. The representation of technology in a curriculum created by science educators may convey that which scientists need to know about technology and not convey what technologists would relay either through content or theory about their own field. How could this be otherwise? Each of us embedded in our own culture realize we cannot create a curriculum for another field; therefore, we ought to realize that we do not have the expertise to incorporate, faithfully, the content of another field. Yet, educators interested in STS often cite the need for a unified body of knowledge that bridges the academic cultures (Bauer, 1990; Cutcliffe, 1990). If we are to do this, we must rely on the members of the different cultures to teach us about their respective fields, both the language and the culture. Our most pressing challenge is to bridge the cultural divide.

Implications for Research

My thesis has been simple. There are two cultures, science and technology, and the members of the two cultures think and speak differently about the curriculum. The two cultures of technology and science have been created and influenced by Western civilization. In application there has been a relationship between technology and science; but in academe, there has been a hierarchical separation of technology and science. That academic separation is becoming more and more artificial as new vistas of research are further blurring the distinctions between technology and science. Based on the way in which society creates and uses technology and science, rejoining the constructs and concepts of technology and science in academe is essential for the students whom we teach.

Now, members of the two cultures must try to communicate with each other in order to learn about each other's cultures and to plan truly integrated technology and science programs for children. This task will not be easy; our languages are quite field specific, and our vocabularies are extensive (Fort, 1993). We must, however, try to begin to communicate with each other in order to make sense of the relationships between technology and science. In order to rejoin technology and science for children we need to rejoin technology and science in academic communities by identifying the way in which we talk about our respective curricula, the beliefs we hold about our fields, the way we structure our curricula, the constructs and concepts we choose to teach, and the strategies we use to convey what we deem important for students to know, do, and believe. Each cultural community has to investigate the other community in order to begin to have enough common knowledge to proceed with a real discussion about the rejoining of technology and science. When we can develop a common language, we can then attack the larger concerns of what we have in common and how we can best organize and teach those commonalities and relationships to children. Then, both communities of educators will be working together as one in order to plan integrated curriculum.

References

American Association for the Advancement of Science. (1989). *Science for all Americans*. Washington, D.C.: Author.

Bauer, H. H. (1990). Barriers against interdisciplinarity: Implications for studies of science, technology, and society (STS). *Science, Technology, and Human Values, 15*(1), 105–119.

Bishop, J. E. (1993). Unnatural selection. *National Forum, 73*(2), 27–29.

Brusic, S. A. (1992). Achieving STS goals through experiential learning. *Theory into Practice, 31*(1), 44–51.

Bybee, R. W. (1991). Science-technology-society in science curriculum: The policy practice gap. *Theory into Practice, 30*(4), 294–302.

Cutcliffe, S. H. (1990). The STS curriculum: What have we learned in twenty years? *Science, Technology, and Human Values, 15*(3), 360–370.

DeVore, P. W. (1980s). *Technology*. Worcester, Mass.: Davis.

Dewey, J. (1925). *Experience and nature*. Chicago, Ill.: Open Court.

Fort, D. C. (1993). Science shy, science savvy, science smart. *Phi Delta Kappan, 74*(9), 684–689.

Fox Keller, E. (1985). *Reflections on gender and science*. New Haven: Yale.

Gaskell, P. J. (1982). Science, technology, and society: Issues for science teachers. *Studies in Science Education, 9*, 33–46.

Gilliom, M. E., Helgeson, S. L., and Zuga, K. F. (1991). This issue. *Theory into Practice, 30*(4), 232–233.

Heath, P. A. (1992). Organizing for STS teaching and learning: The doing of STS. *Theory into Practice, 31*(1), 52–58.

Hickman, L. A. (1990). *John Dewey's pragmatic technology*. Bloomington, Ind.: Indiana University Press.

Hurd, P. DeH. (1991). Closing the educational gaps between science, technology, and society. *Theory into Practice, 30*(4), 251–259.

Ihde, D. (1990). *Technology and the lifeworld.* Bloomington, Ind.: Indiana University Press.

Kevles, D. J. (1993). Social and ethical issues in the human genome project. *National Forum, 73*(2), 18–21.

Kowal, J. (1991). Science, technology, and human values: A curricular approach. *Theory into Practice, 30*(4), 267–272.

Kranzberg, M. (1991). Science-technology-society: It's as easy as XYZ! *Theory into Practice, 30*(4), 234–241.

Lux, D. G. (1981). Reality, Aristotle, and the teaching of learning about and learning how-to. *School Shop, 40,* 24–25.

Lux, D. G. (1984). Science and technology: A new alliance. *The Journal of Epsilon Pi Tau, 10*(1), 16–21.

May, W. T. (1992). What are the subjects of STS—really? *Theory into Practice, 31*(1), 73–83.

Rubba, P. A. (1991). Integrating STS into school science and teacher education: Beyond awareness. *Theory into Practice, 30*(40), 303–308.

Selby, C. C. (1993). Technology: From myths to realities. *Phi Delta Kappan, 74*(9), 684–689.

Snow, C. P. (1959). *The two cultures and the scientific revolution.* New York: Cambridge University Press.

Splittgerber, F. (1991). Science-technology-society themes in social studies: Historical perspectives. *Theory into Practice, 30*(4), 242–250.

Whitehead, A. N. (1925). *Science and the modern world.* New York: Macmillan.

Winner, L. (1977). *Autonomous technology: Technics-out-of-control as a theme in political thought.* Cambridge: MIT Press.

Zuga, K. F. (1991*a*). The technology education experience and what it can contribute to STS. *Theory into Practice, 30*(4), 260–266.

Zuga, K. F. (1991*b*). Technology teacher education curriculum courses. *Journal of Technology Education,* 2(2), 60–72.

PART IV

STS Initiatives Outside the United States

STS is a worldwide reform effort in science education. The current focus on the 2000+ initiative of UNESCO is a perfect example. Technology is emerging as a major focus in schools and one affecting the science curriculum. The difficulties with such a focus for all students demonstrate the distinct dichotomy separating the human-made and natural worlds in school programs. Most now see the union as a necessity. A central focus on technology makes science study more personal, current, and local. Engaging students is essential if learning is to occur, even when it means less concern for natural science per se. More students can see the power of technology—because of its immediate impact on daily living more so than they can of natural science.

To be sure STS initiatives began earlier in some nations. However, with the focus on STS in the United States and the research to illustrate its power, the worldwide interest in STS is intensifying, especially in Asia.

Chapter 21

STS in Britain: Science in a Social Context

Joan Solomon

Prelude to STS

The beginning of STS in Britain could be dated either 1970 or to a progression of different influences felt very much earlier. Science education had been slow to gain a secure place in the humanities-dominated school curriculum of the 1920s and 1930s. When it did succeed, science claimed its place as a rigorous intellectual discipline asserting that it described the processes of nature most rigorously, and yet showing very little connection with everyday life. It had the prestige, and indeed rarity, to insist on the validity of its knowledge with little fear of contradiction.

The year 1970 saw the inception of an association of British science teachers from universities and polytechnics who wanted to make substantial changes in the teaching of science. Their immediate objective was to prepare teaching resources about the philosophical nature and social impact of science. The association took the name SISCON and was led by Dr. Bill Williams of Leeds University.

Although it was new and radical in its thinking, SISCON had been forged out of ideas extending over several decades. The 1930s, for example, were a time when it began to seem that science might have something of special value to offer ordinary citizens, and especially to the poor and oppressed. Left-wing scientists, like the Irish physicist J. D. Bernal, claimed that history showed the force of science and that it was a power for intellectual liberation. Lancelot Hogben, another member of this group of radical scientists, also tried to "bring science to the people." In the introduction to his book *Science for the Citizen*, Hogben wrote that it was intended "for the large and growing number of intelligent adults who realize that the Impact of Science on Society is now the focus of a genuinely constructive social effort" (Hogben, 1938, p. 9). These scientific humanists presented "high science" almost as a substitute morality, in a way, which would be called scientism today. Their immediate effect on school science education was negligible.

World War II had a much stronger influence. The best scientists from all over Europe had flocked to Los Alamos to work, quite knowingly, on what they called "our bomb." The effects of actually dropping it on Japan had shocked them. In postwar Britain new associations of young scientists sprang up that were dedicated to showing they could and should be responsible to society for the artifacts and knowledge of science.

About the same time a very influential lecture was given by Charles Snow on *The Two Cultures* (1959). It argued that the prevailing culture of the arts and humanities was ignoring the importance of scientific knowledge. The debate was prominent for a time; it even found a place in some lighthearted songs of the day. The context of the argument, however, was still academic: it was about the public understanding of the knowledge on thermodynamics versus their understanding on the works of Milton and Shakespeare. Nonetheless, the argument did serve to focus attention on an area that would later become known as public "scientific literacy" and thereby impinge on STS from a different direction.

The classical view of science as the disinterested pursuit of truth remote from society and politics was also being attacked by the philosophy of science. This had begun to look inward, to see the scientific community rather as a social anthropologist might see South Sea islanders. The work of Thomas Kuhn (1962) and John Ziman (1968) explained the construction of scientific knowledge as a social activity in itself, as well as one that affected the quality of life within the wider society.

The new SISCON organization debated these new trends in the sociology and philosophy of science at their annual summer schools. They also wrote materials to incorporate them into their teaching.

STS IN SCHOOLS

The need for a new kind of school science education arose in some measure from an influential report, *The Limits to Growth* (Meadows, et al., 1972), which started a debate about (1) the exponential growth in fuel use and the finite nature of fossil fuel reserves and (2) the world population explosion and its limited production of food.

At first the discussion of such topics was confined to the top grades in the old British public schools which had, inevitably, a rather elitist and technocratic flavor. Nevertheless, it produced the first STS course in Britain with good materials about alternative energy resources. In 1978 a movement in British state schools was started by two teachers who had attended the original SISCON seminars. It was called SISCON-in-Schools and, in one form or another, is still in existence. It was dedicated to a kind of citizen science that saw science-based social issues as an essential part of school science education whose wider task was to prepare students for decision-making in a democratic society that was more and more bedeviled by such problems.

The SISCON school units were developed during the 1980s through a process of action research that first delivered pilot material, and then embarked on a series of emendations and changes as it was tested by an energetic group of London school science teachers. The first materials were designed for older students (age 17), although not those with especially high ability. The order of the eight units was indicative of the intellectual structure of STS studies as the group and its advisers saw it. The first two units were the most theoretical and they set up the general knowledge base for the other six, which were more concerned with areas of particular problemmatique but used the same historical and philosophical approach. The units were:

1. Ways of living (interactions with society)
2. How can we be sure? (nature of scientific knowledge)
3. Technology, invention, and industry
4. Evolution and the human population
5. The atomic bomb
6. Energy: The power to work
7. Health food and population (mainly third world)
8. Space, cosmology, and fiction.

Finally, after the fifth revision, the series was published by the Association for Science Education in 1983, together with a Teachers' Guide. All the materials included historical and philosophical strands as well as an emphasis on care for the environment and technology for the third world. In effect, it had a very broad sweep and there is little doubt that, at the beginning, only the real enthusiasts felt comfortable teaching it. One of the aims of the program was to enable students to consider and discuss controversial issues, and this immediately called into question the normal teacher-led lecture approach to learning. New methods of teaching, such as role-play and small group discussion, were beginning to be advocated, but they were not so easily learned and used by teachers more familiar with the didactic approach to laboratory and theoretical work. Some examples of the more successful strategies were reported in the *Teachers' Guide* (Addinell and Solomon, 1983), but no substantial efforts were made to validate them. This lack was probably due to the practitioner nature of the leadership, which put teaching above research; this leadership also had serious trouble in attracting proper funding for this forbiddingly new venture.

However, by 1984 a syllabus, based on the SISCON materials and ideas, was accepted by one of the most prestigious examination boards in England. The movement was on its way.

In the years that followed another project with quite a different approach to STS began work. This time the underlying intellectual structure of the subject was almost totally ignored. The objective was to prepare a series of curriculum materials that would "fit in" with existing science syllabuses and individual teacher preference. They contained a wide range of approaches from the statistical to the empathic, from simple applications of newly learned science concepts to role-play in a public or industrial setting. These units were written by members of the Science and Technology in Society (SATIS) (Holman, 1988) project, all of whom were also practicing school science teachers. This time, however, the teachers were advised not by STS scholars but by industrialists and practicing scientists. Once again, the materials were tested in schools, and thereby were generally welcomed by science teachers as "user-friendly." They successfully extended STS studies down the age range to fourteen- to sixteen-year-olds and then to nine- to fourteen-year-olds.

By 1985, then, a curious situation had developed in which the STS movement was clearly and prominently in existence, but supported by little literature, with the notable exception of Ziman (1980), and even less research.

About Enjoyment and Achievement in Science

When research on STS first began, it had just two main themes: one was about the greater involvement of girls in science and the other was the changes in attitude and achievement in traditional school science that STS might be able to produce.

The very small number of girls recruited to the physical sciences was a cause for educational concern and an object of considerable research during the early 1980s. Whether these were statistical surveys of the general interests of students (Collings and Smithers, 1984) or psychological explorations of the students' reasons for subject choice (Head, 1985), the results were similar. Girls were more interested in topics related to people than were boys. Several writers on gender were quick to recommend that including the social implications of science would make it more appealing to girls. It was also about this time that the rather patronizing term "girl-friendly science" was first used. This was science connected with those topics that girls were supposed to like: in addition to cooking and cosmetics, some wiser authors began to include social issues like genetic counseling and environmental pollution. A longitudinal study of boys and girls studying physics from ages fourteen to nineteen by Anthony Pell (1985), who was himself one of the early SISCON-in-Schools pioneers, showed that the inclusion of social issues and also the philosophical aspects of science, correlated well with the girls' achievement in examinations and enjoyment of the subject. The additions seemed to have less impact on the boys' enjoyment of physics, which was most significantly linked to "finding it easy."

As interest in teaching relevant science increased, so research on how students reacted to these lessons became a little more common. Some were based on no more than pencil-and-paper evaluation sheets (e.g., Walker, 1990; Lenton, 1991), none of which added to an understanding of how learning through STS affected students. A later course, which incorporated STS, *Salter's Science*, was evaluated by Judith Ramsden (1992) in a deeper sense: she reported the curious finding that although students thoroughly enjoyed the course, some were worried lest it was not "real" science. This gave clear indications of the strength of the public image of science as academic, amoral, and uncaring.

These results had also been found in some action research reported earlier by Solomon (1985). Here the social effects of the energy crisis of the 1970s were discussed by pupils aged fifteen. It was shown that conventional teaching about the physics of energy, including the principle of conservation of energy, actually inhibited the brighter pupils from expressing their opinions on the social uses of energy. As in Ramsden's later study, some pupils were found to ignore questions about social implications included in routine tests on the grounds, it seemed, that they were not part of real physics! This longer-term study also showed pronounced gender effects, just as the earlier studies had. Girls and boys responded on paper in different ways to the science connected to social and political issues: the girls showed more concern and involvement with context, and the boys gave more impersonal scientific comments.

Influences from Outside School

In his seminal book on teaching and learning STS, John Ziman (1980) had attached various aspects of scientism such as the belief "that only the scientific expert can give really reliable advice" (p. 47), or "the tendency of science education to present (scientific knowledge) as value free or morally neutral" (p. 47). This was a challenge to schools to explore students' values as well as teaching scientific knowledge. These two approaches to the social issues of science—the knowledge based, and the value ridden—seemed miles apart, at the very time when public debate about nuclear power and environmental control became ever more heated. In 1985 the Royal Society produced a report called *The Public Understanding of Science* (Bodmer, 1985), which seemed to take the simplistic view that learning more science at school would enable people to understand better the reassurances of the scientific experts. For all aspects of the dissemination of science knowledge, it recommended less "sensationalism" and made no mention of personal values. Nevertheless, the report stimulated a funded research program that valuably covered many fields of public understanding, including school students' discussion.

British science education had gradually moved toward the use of discussion-based approaches to the socioeconomic and applied aspects of science. While this certainly did not take place in every classroom, it was strongly advocated by the Association for Science Education and even supported by comments from the Inspectorate (DES, 1985). The rationale for such changes was child-based, hoping for a better attitude (e.g., Byrne and Johnstone, 1988) derived from the use of more accessible learning methods, rather aiming for a new understanding of the social nature of science. When the first national science curriculum emerged in 1990, it seemed as if these trends might continue and indeed broaden out into a proper exploration of the nature of science. For two years, from 1990 to 1992, research into students' understanding of science was actually funded by the National Curriculum Council (Solomon, et al., 1992), before the curriculum was drastically changed so that it apparently excluded specific study of the nature of science.

Knowledge, Affect, and Scientific Literacy

At the heart of all STS lies the conjunction of the affective and the cognitive, that is, the students' values and their scientific knowledge. The deep research question that cried out for exploration was how these diametrically opposed approaches to the scientific issues worked together. This was no longer a question of devising new teaching strategies, but of exploring the results of them in action.

The three-year DISS project (Solomon, et al., 1992) used a new methodology. It deliberately introduced excerpts from general release television programs into STS lessons in different schools and asked the students to form groups and record their discussion of the issues in the excerpt that they had just viewed as they perceived them. Almost every excerpt involved affective issues (kidney donation, incidence of leukemia, third-world medicine, etc.) as well as referring to scientific knowledge. Pre- and postcourse questionnaires were administered about attitudes, preferred media for access to scientific knowledge, and understanding of the nature of science. The results, from both the discussions and the questionnaire, were too rich to report fully here. However, they are very germane to STS teaching and learning as well as to the Public Understanding of Science. The results are summarized as follows:

1. Familiarity with scientific terms and concepts, rather than deep knowledge, was an essential prerequisite for attending to science-based social issues.
2. Contextual and affective discussion always preceded group decisions on strategies for delivering social justice.
3. Although boys were more likely than girls to begin discussions with broad judgmental comments, during most of the discussion the genders exchanged ideas and arguments in similar ways.
4. In the questionnaire boys and girls presented themselves and their preferences quite differently. These were often contradicted by their performance during the discussions.
5. Very valuable short- and long-term reflective effects were produced by taking part in these discussions.

It would be pleasant to be able to finish this chapter about British STS on an optimistic note, but the present educational climate of rigorous testing and itemized knowledge promises little. However, STS teaching did gain a sturdy foothold in the British science curriculum during the 1980s and it is to be hoped that this will not completely give way under present pressures.

References

Addinell, S., and Solomon, J. (1983). *Teachers' guide to the SISCON-in-schools project.* Hatfield, England: Association for Science Education.

Bodmer, W. (1985). *The public understanding of science.* London: Royal Society.

Byrne, M., and Johnstone, A. (1988). How to make science relevant. *School Science Review, 70*(251), 43–46.

Collings, J., and Smithers, A. (1984). Person orientation and science choice. *European Journal of Science Education, 6*(1), 55–65.

Department of Education and Science. (1985). *Science 5–16: A statement of policy.* London: HMSO.

Head, J. (1985). *The personal response to science.* Cambridge, England: Cambridge University Press.

Hogben, L. (1938). *Science for the citizen.* Woking, England: Allen and Unwin.

Holman, J. (1988). Science-technology-society education. *International Journal of Science Education, 10*(4), 343–345.

Kuhn, T. S. (1962). *The structure of scientific revolutions.* London: University of Chicago Press.

Lenton, G. (1991). A review of SATIS national trials. *School Science Review*, *73*(262), 7–18.

Meadows, D., et al. (1972). *The limits to growth.* Washington, D.C.: Potomac Associates.

Pell, A. (1985). Enjoyment and attainment in secondary school science. *British Educational Research Journal*, 22(2), 123–132.

Ramsden, J. (1992). If it's enjoyable is it science? *School Science Review*, *73*(265), 65–71.

Snow, C. (1959). *The two cultures* (original work). Published (1965) as The Rede Lecture. Cambridge, England: Cambridge University Press.

Solomon, J. (1985). Learning and evaluation: A study of school children's views on the social uses of energy. *Social Studies of Science*, *15*, 343–371.

Solomon, J., Duveen, J., Scott, L., and McCarthy, S. (1992). Teaching about the nature of science through history: Action research in the classroom. *Journal of Research in Science Teaching*, *29*(4), 409–421.

Walker, D. (1991). The evaluation of SATIS. *School Science Review*, *72*(259), 31–39.

Ziman, J. (1968). *Public knowledge*. Cambridge, England: Cambridge University Press.

Ziman, J. (1980). *Teaching and learning about science and society*. Cambridge, England: Cambridge University Press.

CHAPTER 22

STS THROUGH PHYSICS AND ENVIRONMENTAL EDUCATION IN THE NETHERLANDS

Harrie M. C. Eijkelhof
Koos Kortland
Piet L. Lijnse

STS IN PHYSICS EDUCATION

In the Netherlands STS initiatives date back to the end of the 1960s. This chapter describes the origin of these initiatives and outlines some influential projects in the field of STS for secondary education in which the authors have been involved. The chapter also presents a reflection on STS teaching from a research point of view. The reader should bear in mind that in secondary schools in the Netherlands the sciences are offered as separate subjects.

Like in many other countries, movements toward STS in the Netherlands originated at the university level. Toward the end of the 1960s, the role of science and technology in society among students and staff was a major topic of discussion, principally related to the nuclear arms race, to the Vietnam War, and to environmental problems. At almost all Dutch universities STS courses were developed. In 1978 and 1982, international conferences were held about STS in Amsterdam (Boeker and Gibbons, 1978) and Leusden (Slaa, Turkenburg, and Williams, 1983) in which the main participants were university staff from the United Kingdom and the Netherlands. Some participants also came from the field of secondary science education, but they were a small minority.

The initiators of these two conferences had started in 1976 at the Free University of Amsterdam with the development of a course called physics in society. It was to be used as an optional unit in the Dutch physics education curriculum for preuniversity pupils. This course was later published in an English version (Eijkelhof, Boeker, Raat, and Wijnbeek, 1981). The book deals with complex socioscientific issues like energy production and consumption, noise, traffic and nuclear arms, and with reflection on the role of science and technology in developing countries and in industrialized countries. The "course" was small (it covered only fifteen lessons) and could be used as a supplement to any physics textbook, which in those days paid little attention to the interaction between physics, technology, and society. The course appeared to be very popular among teachers and students (Eijkelhof and Swager, 1984).

Another sign of the increase in interest in STS among teachers in the Netherlands was a well attended summer conference about physics and society teaching organized by the Dutch Association for Science Education at the Free University of Amsterdam in 1979, in which teachers from some British projects acted as invited speakers.

In the beginning of the 1980s the STS challenge was taken up by the PLON-project. This physics curriculum development project (Eijkelhof and Kortland, 1988) had started in 1972 with the development of new materials for physics teaching, initially for junior and later for senior secondary education. During the first eight years of the project, its aims were mainly to make physics education more attractive to students by relating physics to daily life phenomena, by allowing students to make choices in subtopics to be studied, and by using a large variety of working methods, giving special attention to group work and to presentations about the work done in these groups.

From 1980s onward the PLON-project developed four optional STS units for junior secondary education. These units focused on specific issues such as the use of alternative energy sources, the nuclear arms race (Eijkelhof, et al., 1984), the use of water pumps in a developing country (PLON, 1985), and traffic problems. Unfortunately, these units were not used widely, partly because of the pressure from the rather full obligatory part of the curriculum and partly because physics teachers did not feel at ease with teaching units in which nonscientific aspects played such an important role.

At the senior level it was decided not to develop specific STS units but to incorporate STS aspects at numerous places in all units (Eijkelhof and Kortland, 1988; Lijnse, et al., 1990). Some examples of STS aspects in units are:

Units	*STS Aspects*
Traffic and Safety	Fuel Conservation, Traffic Safety Device
Matter	Social and Scientific Contexts of Fundamental Research
Ionizing Radiation	Risk and Safety Aspects in a Variety of Applications (PLON, 1984)
Satellites	Current Applications of Satellites for Communication, and Earth and Astronomical Observations

Aims related to these STS aspects were:

1. the use of knowledge and skills in out-of-school situations, that is, in selecting and using products, and in understanding socioscientific issues; and
2. acquiring a more authentic image of the work of physicists in the past and the present.

Partly due to experiences with these units, a considerable number of these STS aspects were incorporated into the syllabi for the new national examination for physics in senior high schools (effective from 1993 onward). As a consequence many of the new physics textbooks gave more attention to STS aspects than the previous generation of physics textbooks, although none of them went as far as the PLON units.

Environmental Education

In 1986 curriculum development started in the area of environmental education. Funding from the national government became available, mainly as a result of a growing awareness that measures counteracting the trend of environmental deterioration would adversely affect society. The concerns of the government about public acceptance of these measures made officials think of incorporating environmental education into the curricula of a number of existing school subjects. The NME-VO project was initiated to accomplish this. In this project science educators and nongovernmental environmental organizations cooperated to produce a number of exemplary teaching units for biology, geography, physics, and chemistry courses.

The teaching materials were to be based on *science-and-technology-related social (environmental) issues*, with (traditional and new) content and skills from the different school subjects serving as tools for *decision-making* on these issues and their implications in everyday life (Kortland, 1987). With regard to the contribution of the different school subjects, a choice for *coordination* was made, instead of a complete integration—thus complying to the Dutch tradition of teaching the sciences as separate subjects. The project developed a number of teaching units for different age-groups, ability-levels, and school subject combinations. The issues addressed were: the tension between human use of wetlands and animal life there, the environmental effects of meat production, burning fuel, traffic noise, pesticides, and waste.

A teaching unit starts with an orientation on the issue, leading up to a central question setting the scene for the rest of the lessons. This orientation is followed by two or more different parts, one for each of the participating school subjects. These parts of the unit focus on an analysis of the issue, with an introduction of the relevant disciplinary contents and skills. In the next part of the unit alternative solutions to the issue are investigated and assessed by different groups of students. The final part of the unit deals with weighing the alternative solutions: What seems to be the best one from environmental and other (economic, social) points of view? A teaching unit as a whole reflects some kind of a *structured decision-making process*: identifying a problem, developing criteria, generating and evaluating alternatives, and choosing the (best) solution (Kortland, 1989). For students who might want to go further, most of the issues chosen provide possibilities for action-taking in everyday life.

An Example: Fuel

The chemistry/physics/geography-unit *Fuel* (grade 10, high-ability level) relates to the regular program in these school subjects: burning in chemistry, energy in physics, and spatial planning in geography. The topics dealt with in the unit are interrelated, but the structure of the unit is chosen in such a way that for the three school subjects "only" a coordination in time is required. Chemistry starts by giving an orientation on the energy issue and proceeds with burning of fossil fuels: reaction products, environmental effects, countermeasures (using alternative fuels and ways of burning). At a point in time in physics topics, such as exhaustion of fuels, the energy system of a house, and ways to reduce the use of fuels (decreasing energy-demand, increasing energy-conversion efficiency, and using renewable energy sources) are introduced. This is the "background information" students take with them to the geography lessons where they learn about spatial planning—and then student-groups apply their knowledge and insight to decide about an appropriate combination of energy system and spatial plan for a new housing estate. Finally (in the geography lessons), the different plans for the new housing development are exchanged, discussed, and assessed on aspects like the level of fuel consumption, observance of spatial planning rules, attractiveness of the area for living.

These teaching materials have not yet been widely used, first of all because they require additional costs for the schools and because they introduce several educational innovations, including an issue-based approach, new content and skills, decision-making, and school subject coordination. It is especially this last point that causes some difficulties in trial schools, even though these schools had volunteered for the project because of the prospect of school subject coordination or even integration.

In the meanwhile, preparations for a reform of school curricula for junior secondary education had started. Among other changes, environmental education—more or less based on the experiences of the NME-VO project—became an explicit part of the curriculum for a number of school subjects: geography, history, economics, technology, physics, chemistry, and biology. In physics, chemistry, and biology, science-and-technology-related social issues regarding energy, waste, noise, and food are expected to be dealt with in the context of decision-making by students, indicated by attainment targets like "students are able to present an argumented point of view in a situation of choice."

The large number of school subjects of which environmental education was to become a part called for some coordination. Therefore, a core curriculum (Pieters, 1990) was developed, providing a common set of goals and conceptual structure for further elaboration within the curricula of the different school subjects. Empirical foundation of these curriculum documents started in 1990 by means of a number of case studies of classroom practice during trials of teaching units developed for research purposes (Kortland, 1992). Some results of this research will be dealt with in the next section. However, this combination of curriculum development and in-depth research was terminated in an untimely manner at the end of 1992, due to a change in the government's funding priorities in this area. Most of the money available is now directed at the introduction of environmental care systems at school level and at large-scale implementation of environmental education at the classroom level. Teachers are expected to develop and implement their own type of environmental education by networking with colleagues.

Research and Reflection on STS

The above mentioned projects imply a broadening of aims, at the same time trying to improve science teaching and learning. As such, STS is but another innovation. Since the fifties, a large number of curriculum development projects have been developed from several different perspectives. Emphasis has been on teaching "the structure of the discipline"; on "being a scientist for the day," and "discovery learning"; on Piagetian theory and stages of cognitive development; and since the eighties on STS. One may rightfully ask In what way have all these efforts really resulted in "better" science education? Why would STS succeed, where other attempts have largely failed? In fact, little is known about the success of STS in reaching the aims that are claimed. We can only critically reflect on STS, based on our experiences and research, related to the implementation and evaluation of the above-described projects.

The Dutch PLON and NME-VO projects have made quite an effort to develop STS curricula at the secondary level; their rationale can be briefly characterized as "teaching science actively in everyday and issue-related contexts." From such teaching, it is expected, on the one hand, that students will find the content taught as more relevant. On the other hand, it can be argued that they will be better able to understand and connect the concepts and skills learned to their out-of-school world. However, how realistic are these expectations?

Before we report some research results, we first want to point out a problem that may be special for the Dutch educational setting. As previously described, the PLON curriculum is first of all a physics curriculum. STS aspects were integrated while teaching "real" physics. This has a clear advantage, but at the same time an essential tension results from it. The criterion for the inclusion of concepts and their level of development was not set purely from an STS point of view, but predominantly from the point of view of regular physics teaching. This caused an unclear duality in the curriculum construction and in its implementation. For teachers, however, the situation was clear; students have to be able to pass the regular final physics examination. That meant that in case of doubt or of if there were time constraints, the STS aspects were limited or nonexistent.

In the case of the NME-VO-project, the situation is almost the reverse. The NME-units could be designed completely from a particular STS perspective. However, the problem was that environmental education is not a recognized school subject. Therefore, the compromise of inclusion in several well-established school subjects had to be chosen, which caused problems of coordination between subject teachers.

In both cases, however, the setting is certainly not optimal in order to achieve the STS aims to be pursued and to collect evidence of such achievement.

On the one hand, extensive evaluation of parts of the PLON curriculum (Wierstra, 1990) has shown that in general students do indeed find the PLON curriculum as more related to their lives beyond school and consequently as more relevant and motivating than traditional curricula. Nonetheless, the differences were less than hoped for and expected. The most positive was the fact that students appeared to like the active classroom climate that is typical for PLON-lessons. However, for both factors, that is, content related to the real world and student centeredness, the personal influence of the teacher seems to be at least as important as that of the curriculum content (Brekelmans, 1989).

On the other hand, the cognitive learning outcomes of the PLON-curriculum, in general, were not significantly different from those of traditional curricula. This unexpected outcome provided a reason for further reflection and in-depth follow-up studies.

Our experience with a unit on *Ionizing Radiation* (Eijkelhof, 1990; PLON, 1984) for senior high school (higher ability) students, may serve as an illustration. Teaching about ionizing radiation and radioactivity in the context of risk and risk perception can be regarded as a prototypical example of STS teaching. A summative evaluation showed that though students valued the topic (particularly girls), cognitive learning results were disappointing. In particular, students were not inclined to use the knowledge taught in their reasoning about situations for which they already had an opinion, such as: disposal of nuclear waste, nuclear energy, having an X-ray. At the same time, in their common sense reasoning about familiar situations students appeared to use a number of "lay-ideas" about radioactivity, which were in conflict with the knowledge taught.

Both situations indicate that using the knowledge of science taught in everyday situations is much less self-evident than was expected. A transfer step is needed that is usually not made if students have already developed familiar and satisfactory common-sense explanations available. This conclusion is endorsed by the work of Van Genderen (1989) on the learning of mechanics in an instructional unit on "Traffic," and of Van der Valk (1992) in his research on the learning of energy in the context of the social problem of energy supply. Kortland (1992), in his work on "Waste," showed that for younger students the aim of using taught knowledge in a rational decision-making process is very difficult to achieve. This difficulty is not only related to the knowledge to be used, but particularly also to the question of what constitutes a rational way of making decisions. From their common-sense point of view, students often see, for example, no need to use extensive comparisons of disadvantages and advantages to come to a decision.

In fact, we are dealing with the same problem in three different disguises. In the learning of concepts, in using concepts in the formation of an opinion, and in weighing opinions in the process of decision-making, we encounter the problem that for many situations students use common-sense concepts to express common-sense opinions, possibly in a common sense process of decision-making. The need to replace such common-sense patterns of reasoning by more scientific patterns is not at all self-evident for students.

Now, in retrospect, these learning and teaching problems are to a large extent understandable. In the last decade, studies on teaching and learning have emphasized the importance of what has become known as "alternative frameworks" that students bring to the classroom. This may explain why past curriculum efforts have been only moderately successful. Does this also apply to STS? Therefore, it is necessary to reflect on the teaching strategy used in most STS curricula of the past, including the projects described above.

In general, one could say that in a traditional "structure-of-the-discipline" type of curriculum, the concepts to be taught are the basic concepts of science. The sequence in which they are taught reflects its basic "logical" structure. The situations in which these concepts are to be used are the usual idealized "scientific" situations. It is precisely this latter aspect that is an important consideration when with the essence of what many STS curricula want to change. But teaching in real life contexts leaves the conceptual structure and sequencing essentially unchanged the value of the context may be lost. Apart from the fact that, because of the complexity of real situations, some new concepts may have to be added (see e.g., De Jong, et al., 1990).

We characterize such a teaching strategy as "top-down," that is, both in traditional and in the PLON curricula (as well as in many other STS curricula) teaching starts from and focuses on the perspective of science; teaching it without really taking into account what students already know and think (see below). Essentially, it is aimed at a direct "top-down" transmission of concepts, even though the way in which that is done may include many open discussions and discovery activities. In both types of curricula, such teaching unavoidably results in a process of "forced" concept development, which may explain the apparent lack of differences in the learning of specific concepts mentioned above.

However, as already noted, research has shown that students' alternative frameworks have to be taken into account, asking for, as it is argued, conceptual change instead of conceptual transmission. As described, common-sense ideas in STS teaching play an even more unavoidable role than in traditional teaching. Students not only appear to have common-sense ideas related to the concepts to be taught, that can certainly not be avoided while dealing with real-life contexts from which they stem, but also opinions and knowledge related to these contexts themselves. They also have common-sense ways of reasoning about and making decisions in such contexts. It is this latter type of preknowledge that explains why, even if we would succeed reasonably well in teaching correct conceptual knowledge, it may still not be used in real-life situations, as we have described. It would not result in an integration of science concepts with ways of reasoning in and about contexts, but in science concepts as something that is added on.

We may conclude that the importance and analysis of common-sense thinking and reasoning patterns have been underestimated in many STS curricula, as well as their difference with "scientific" reasoning patterns. This results in an overestimation of the importance of the latter for one's everyday life and an underestimation of teaching and learning problems.

CONCEPTUAL CHANGE AS IMPROVED "TOP-DOWN" TEACHING?

Thus, it was concluded that we should try to find a way to improve our teaching strategies, by taking into account students' common sense knowledge and reasoning patterns. This reflects our adoption of a "constructivist" perspective, which means that we agree with statements like "meaning is constructed" and "concepts cannot be transferred from teachers to students" (Duit, Goldberg, and Niedderer, 1992). As such, this is an important nontrivial change of perspective. However, adopting a "constructivist" perspective does not yet say very much about how to teach. The phrase "the teacher must have a good idea of what concepts the students might already have and then engage students in activities that would help them construct the desired understanding" (Duit, Goldberg, and Niedderer, 1992) may easily be interpreted in an insufficient way.

Freudenthal (1991), in a comment on "constructivism," writes as follows: "If I were to accept the term 'constructivism,' I would mean a programme having a philosophy that grants learners the freedom of their own activity."

In our opinion, he points to a basic problem. When this freedom of learners is ignored, teaching unavoidably results in "forced" concept development and thus very probably in misconceptions. It is almost a contradiction to adopt "constructivism," that is, the view that students construct their own meaning based on what they already know, and at the same time either to prescribe what they have to construct or to devaluate immediately what has been constructed. The basic problem for constructivist teaching thus is how to design teaching such that it guides students to *construct in freedom* the very ideas that one wants to teach. Freudenthal calls this learning process "guided reinvention." In most "constructivist models of teaching" so far worked out, it is precisely this necessary freedom of learners to make and follow their own constructions that is either lacking or being underestimated. In fact, one could then cast reasonable doubt on whether such approaches should be called constructivist at all. For instance, in the status-changing model of conceptual change (Posner, et al., 1982), students' conceptions are essentially considered as wrong ideas that have to be changed as quickly as possible. To do so, the teacher should design activities that lower the status of students' ideas and raise the status of ideas taught. It is hard to see how such an approach may build positively on students' own constructions. This also applies, to a lesser or larger extent, to conflict-strategies (Nussbaum and Novick, 1982) or to the Children's Learning in Science Project (CLISP)-approach as described by Driver and Oldham (1986). So we do not agree with Scott et al. (1994) when, in describing several routes to deal with preknowledge, they say: "Each of these routes attempts to make links between students' thinking and the science view and might therefore be considered to be equally valid constructivist teaching approaches." It is the way in which that link is being developed that makes a crucial difference. Otherwise, the term "constructivist teaching" becomes almost meaningless.

Of course, this is not to say that status-changing or conflict approaches may not improve the learning results as compared to those of traditional teaching. It does explain, however, that the scope of such improvements is and will remain limited. Basically, these approaches could be characterized as using new strategies to improve top-down teaching.

A "Bottom-Up" Perspective for STS Education?

In our opinion, a more radical change seems to be needed, particularly for STS. The aims of STS ask that students are able to use scientific knowledge in their reasoning about out-of-school contexts, or even to reflect on the social value of such knowledge. In view of the above, one cannot but conclude that such aims seem to be rather pretentious. A necessary prerequisite is that students really understand what they are learning and are able to integrate it constantly in their "living use of language." This probably asks for a bottom-up learning process. In analogy to Freudenthal, we could say that we should not teach the concepts of science (as a product), not even in real-life contexts and/or in an above-mentioned constructivist way, but guide students in the activity of "scientificalizing" their world. Insightful learning can only start where students are, which asks for understanding the concepts, basic ways of reasoning, and language they use and motivations they have. Teaching should be carefully designed so that students can build positively on and extend what they already know, in relation to contexts that are meaningful to them. Only then may we expect that knowledge, language, opinions, and actions remain integrated, as STS in particular seems to ask.

We can give no general rules for the design of such teaching. It necessarily asks for an empirical process of closely interconnected research and development. International cooperation in a program of such "developmental research" could probably result in some real progress in our field.

References

Boeker, E., and Gibbons, M. (eds.). (1978). *Science, society and education*. Amsterdam: VU-Bookshop.

Brekelmans, J. M. G. (1989). *Interpersonal behaviour of teachers in the classroom* (in Dutch with an English summary). Utrecht, The Netherlands: W.C.C.

De Jong, E., Armitage, F., Brown, M., Butler. P., and Hayes, J. (1990). *Physics in context*. Melbourne, Australia: Heinemann Educational.

Driver, R., and Oldham, V. (1986). A constructivist approach to curriculum development in science. *Studies in Science Education*, *13*, 105–122.

Duit, R., Goldberg, F., and Niedderer, H. (eds.) (1992). *Research in physics learning: Theoretical issues and empirical studies*. Kiel, Germany: IPN.

Eijkelhof, H. M. C. (1990). *Radiation and risk in physics education*. Utrecht, The Netherlands: CD-Press.

Eijkelhof, H. M. C., Boeker, E., Raat, J. H., and Wijnbeek, N. J. (1981). *Physics in society*. Amsterdam, The Netherlands: VU-Bookshop.

Eijkelhof, H. M. C, and Kortland, K. (1988). Broadening the aims of physics education. In P. J. Fensham (ed.), *Development and dilemmas in science education* (pp. 282–305). London: Falmer Press.

Eijkelhof, H. M. C., Kortland, K., and Van der Loo, F. (1984). Nuclear weapons: A suitable topic for the classroom? *Physics Education*, *19*, 11–15.

Eijkelhof, H. M. C., and Swager, J. (1984). Physics in society. In E. J. Wenham (ed.), *New trends in physics education*, Vol. 4 (pp. 351–359). Paris: UNESCO.

Freudenthal, H. (1991). *Revisiting mathematics education*. Dordrecht, The Netherlands: Kluwer.

Kortland, J. (1987). Environmental education from the perspective of broadening the aims of science education. In K. Riquarts (ed.), *Science and technology education and the quality of life*, Vol. 2 (pp. 546–555). Kiel, Germany: IPN.

Kortland, J. (1989). Environmental education within the science subjects in secondary education: Why, what and what for? AIF/ICASE, Supplemento speziale a *La Fisica nella Scuola*, *22*, 4, 89–106.

Kortland, J. (1992). Environmental education: Sustainable development and decision-making. In R. E. Yager (ed.), *The status of science-technology-society reform efforts around the world, ICASE Yearbook 1992* (pp. 32–39). Hong Kong: International Council of Associations for Science Education.

Lijnse, P. L., Kortland, K., Eijkelhof, H. M. C., Van Genderen, D., and Hooymayers, H. P. (1990). A thematic physics curriculum: A balance between contradictory curriculum forces. *Science Education*, *74*(1), 95–103.

Nussbaum, J., Novick, S. (1982). Alternative frameworks, conceptual conflict and accommodation: Toward a principled teaching strategy. *Instructional Science*, *11*, 183–200.

Pieters, M. (ed.) (1990). *Teaching for sustainable development*. Enschede (NL): Institute for Curriculum Development (SLO).

PLON (1984). *Ioniserende Straling*. Utrecht, The Netherlands: Rijksuniversiteit Utrecht, Vakgroep Natuurkunde-Didactiek. English version (1988): *Ionizing Radiation*. Melbourne, Australia: Monash University, Faculty of Education.

PLON (1985). *Water for Tanzania* (English version). Utrecht, The Netherlands: Rijksuniversiteit Utrecht, Vakgroep Natuurkunde-Didactiek.

Posner, G. J., Strike, K. A., Hewson, P. W., and Gertzog, W. A. (1982). Accommodation of a scientific conception: Toward a theory of conceptual change. *Science Education*, *66*(2), 211– 227.

Scott, P., Asoko, H., Driver, R., and Emberton, J. (1994). Working from chidren's ideas: Planning and teaching a chemistry topic from a constructivist perspective. In P. F. Fensham, R. F. Gunstone, and R. T. White (eds.), *The content of science: A constructivist approach to its teaching and learning* (pp. 201–220). London: Falmer Press.

Slaa, P., Turkenburg, W. C., Williams, W. (1983). *Risk and participation.* Amsterdam, The Netherlands: VU Bookshop.

Van der Valk, A. E. (1992). *Developments in energy education* (in Dutch with an English summary). Utrecht, The Netherlands: CD-Press.

Van Genderen, D. (1989). *Mechanics teaching in motion* (in Dutch with an English summary). Utrecht, The Netherlands: W.C.C.

Wierstra, R. F. A. (1990). *Teaching physics between the daily life world of pupils and the world of theoretical concepts* (in Dutch with an English summary). Utrecht, The Netherlands: CD-Press.

CHAPTER 23

STS INITIATIVES IN JAPAN: POISED FOR A FORWARD LEAP

Namio Nagasu
Yoshisuke Kumano

In the 1980s some Japanese researchers analyzed the SISCON Project in the United Kingdom (Morohashi, 1983) and STS initiatives in the United States, especially those centered in Iowa (Nagasu, 1987). Other Japanese researchers who specialize in the history of science translated Ziman's book, *Teaching and Learning about Science and Society*, into Japanese (Takeuchi and Nakajima, 1989). In addition, one researcher, together with his undergraduate students, developed an STS module based on civil education (Ogawa, 1989).

In the early 1990s, more science educators and science teachers have been interested in STS based on some points of view advanced by leaders in environmental education (Kumano, 1991; Suzuki & Harada, 1990). Other related efforts have focused on the history of science (Ohsu, 1991), the analysis of the Science and Technology in Society project in the United Kingdom (Kurioka and Nogami, 1992; Watanabe and Ikeda, 1992), the analysis of STS in practice and the research base in the United States (Nagasu, 1991), and a national survey of Japanese Science Teachers' Perception of STS Topics (Tanzawa, 1992).

Nagasu et al. (1992) first proposed the possibility of developing STS in Japan at the international conference associated with the Seventeenth Pacific Science Congress STS Session in Hawaii, in 1991.

Some Japanese science educators and other researchers in the 1980s and early 1990s studied the several STS Projects in the United Kingdom and the United States as mentioned above. In other words, there has been a tendency in Japan to investigate and analyze STS as it has evolved in Western countries. In terms of interest in STS, most Japanese science educators and science teachers have focused on STS as one kind of teaching content: for example, the problems and issues of the environment, including water and air pollution, biotechnology, energy depletion, or nuclear fuel. Many Japanese science educators and science teachers have assumed that these problems and issues can be appropriately taught by merely changing course content. They assume further that students will understand the interactions and interrelationships of science, technology, and society simply by studying new, different, and more appropriate material. However, STS means not only new teaching content but the implementation of new theories of teaching and learning. It means evaluation based on new objectives and philosophy. Other considerations of STS in Japan have focused on the goals and approach characteristic of environmental education and its intended relationship to technology education.

STS Ties to Environmental Education and Technology Education in Japan

The latest Revised Course of Study, the Japanese standard curriculum, has recently been approved by the Japanese Ministry of Education, Science and Culture. This calls for school science in grades three through twelve; that is, elementary science in elementary school (3–6) by 1992, general science in lower secondary school (7–9) by 1993, and thirteen science subjects such as unified science, biology, chemistry, geology, and physics which respectively include various applied science subjects as well as pure science subjects by 1994. In addition, there will be special school subjects in technology that focus on industrial arts and homemaking offered only at the secondary level. These subjects include wood-manufacturing, electronics, metal-manufacturing, machinery, cultivation, and computer information.

There are no special technology subjects in Japanese grades one through twelve that include the teaching content of new technologies such as high electronic technology or biotechnology. Nor are special classes in environmental education found in grades one through twelve. Other courses of study are found in the Japanese system for each subject, including Japanese language, arithmetic and mathematics, science, and social studies in the elementary as well as the upper-secondary school. Other subjects such as English, fine arts, industrial arts and homemaking, health and physical education, and music are also included in the secondary school. The U.S. education system has been in use since the occupied era after World War II. Because of this history and this situation, it is difficult to adopt or develop new subjects within the Japanese system for schooling.

Many people in Japan have argued and agreed that environmental education should be taught. However, these proponents have not yet agreed on what kinds of content are appropriate. First, some claim that environmental education should be taught as a new integrated subject consisting of science, social studies, mathematics, language, drawing and handicrafts, fine arts, and moral and ethics education. But, some believe that environmental education should be taught independently. Some insist that if one teaches science or social studies appropriately, then environmental education will also be taught.

Recently, the Ministry of Education (1991, 1993) developed two new teacher's guides for environmental education. One of these was for the elementary level and another for the secondary level. In this new proposal the following are included at both levels: science, social studies, music, moral education, and Japanese language. Arithmetic, homemaking, and drawing and handicraft are included at the elementary level, whereas mathematics, industrial arts and homemaking, English language, and health and physical education are included at the secondary level. An extracurricular program is also included at both levels. In short, the Ministry of Education has decreed that environmental education must be taught independently as a part of the listed subjects and/or extracurricular areas described above. However, the Ministry of Education must develop a curriculum framework based on the approved Course of Study.

There is not much argument about teaching technology in elementary and secondary schools probably because of the general focus on high technology and biotechnology in our society. Nonetheless, some Japanese science educators and secondary science teachers have argued forcefully that these new technologies should be taught in school science.

Arguments about the specific principles which should be taught in environmental education and technology education are advanced. However, more discussion must occur as ties between the environment and technology education are noted. Such discussion of these problems should consider the following:

1. Both environmental and technology education share the same principles and objectives, that is, to develop citizens who will be able to make decisions or solve the problems in their own lives.
2. Both areas are faced with the interactions and interrelations of science, technology, and society.
3. The traditional subjects in science and technology programs in Japan cannot contain nor deal with the additional areas, especially the focus on current problems and the integration of science, technology, and the environment.

These considerations are the same ones providing philosophical and background information for STS. Some see STS as a force for combining all the concerns and individual ideas advanced as correctives.

Extent of Current STS Activities in Japan

In Japan, there are two major associations for science education. One is the Society of Japan Science Teaching (SJST), which mainly consists of science teachers at elementary- and secondary-level and college science educators. It was established in 1952. SJST publishes *Science Education Monthly* and the *Bulletin of Society of Japan Science Teaching*, which is published three times a year. The other major science education association in Japan is the Japan Society of Science Education (JSSE), which consists of scientists, psychologists, educational technologists, mathematicians, science educators, and science teachers. It was established in 1977. JSSE publishes *Journal of Science Education in Japan* three times a year; two publications are in Japanese and one in English.

In 1992, the JSSE held a special session on STS at its National Convention held at Jyouetsu University of Education. Likewise, the SJST established a new subject area, namely, STS/Environment Education at its National Convention in August of 1993 held at Fukuoka University of Education. These sessions at both meetings were well attended and resulted in many intriguing questions and answers by Japanese science educators and science teachers. *Heredity*, which is a very popular magazine with an emphasis on biology education for secondary- and college-level instructors, offered two special issues focusing on STS in biology education in November 1992 to May 1993.

In addition to these developments, Nagasu (1991) and Umeno (1992) separately published STS research reports on secondary education. Both reports were funded by the Ministry of Education. Both reports investigated the features of STS as reform in the United States. The later report investigated SATIS in the United Kingdom and analyzed the relationships between the STS approach and the Revised Course of Study. Moreover, some secondary biology teachers tried to develop STS modules or topics from these reports for use in their schools.

In 1993, Robert E. Yager, an STS advocate from the United States, visited some Japanese elementary and secondary schools, Shizuoka University, and the University of Tsukuba. He offered special lectures for students and college teachers in June. He also actively discussed the STS approach with Japanese science educators and science teachers in Tokyo. He and the participating Japanese researchers, teachers, and administrators positively and enthusiastically talked about STS's historical, philosophical, and practical points of view. As a result of this visit the participants heard and interacted directly with the features of the STS approach in the United States STS was seen as a reform of science education, that is, as a new paradigm shift for science education.

There is a tendency among Japanese science educators to embrace STS as an emerging field of special value for reform in Japan. In fact, there is evidence that the number of investigations dealing with STS increased in just a few years.

Kumano (1993) completed a first careful study of STS teaching in Japan. Data were collected from grade-ten earth science classes at Meikei High School (Tsukuba, Japan) during the 1991–1992 school year and grade-eight science II classes in the same school during 1992–1993. The data consisted of video recordings for two teachers who were teaching at grade eight. In all cases the teacher used the STS approach with some classes and traditional approaches in other classes.

An STS rubric created by Kellerman and Varrella (1993) allowed a classification of the teacher as novice, capable, experienced, or distinguished. This rubric was based on and used for reports on actual visits to STS classrooms and teachers involved in the Iowa-Scope, Sequence, and Coordination Project. Another scale was used to indicate the use of student-centered behaviors to describe the STS approach. It was possible to provide a profile of behaviors for the Japanese teachers. The results of analyzing video recordings in terms of observed teacher behaviors as well as the level and quantity of STS attributes in evidence were compared with similar teachers using STS and traditional approaches in experimental situations in the United States.

Differences in behaviors between STS and traditional classrooms were studied. There were significant differences between the STS and traditional sections in Japan, including: (1) less teacher direction was found in STS sections; (2) more student-centered activities were found in STS sections, especially greater use of cooperative learning; (3) teachers provided more wait time in STS sections and thereby involved more students; and (4) teachers using the STS approach showed more affirmation statements, more caring, more humor, and more encouragement.

The difference in results with student achievement in STS and traditional classrooms was not dramatic. The failure to prepare a reliable and valid instrument in the application domain meant no comparisons could be made. Specific results with STS instruction compared to traditional methods in Japan permit the following statements:

1. STS students do as well as students in traditional classes with respect to concept mastery;
2. STS students in the grade-ten sample developed significantly more positive attitudes about science study, science teachers, and science careers than students in traditional sections;
3. STS students showed significantly more growth with respect to creativity skills in both grades eight and ten; and
4. STS teaching in Japan is underway with the initial results most encouraging.

Poised for a Forward Leap: The STS Approach

The senior author surveyed the STS approach in the United States during the 1991–1992 school year. He clarified the picture of the STS approach by comparing and contrasting it with the so-called Golden Age of 1955–1974 (Nagasu, 1993). He argued that the philosophical features of the STS approach, if implemented, will amount to a new paradigm shift for science education.

A paradigm shift means a change so great that all assumptions and frames of reference change. In Japan this means new goals for science education based on the STS approach. These include:

1. developing and creating a scientific and technological literate citizenry that will be able to solve the new problems and issues or to make responsible decisions after graduating from high school;
2. developing a more aware citizenry than in other industrial countries;
3. developing greater understanding of the new philosophy of science, that is, to observe and make judgments based on value-laden or theory-laden constructs;
4. implementing and using a new theory of learning in science, that is, the constructivist learning theory;
5. utilizing new theories of curriculum making, including interdisciplinary and learner-centered curricula; and
6. developing and using new theories of evaluation, including authentic forms that are tied to instruction and real-world problems and alternative forms based on the constructivist learning theory and actual performance of students in ways other than recalling information for excelling on multiple-choice examinations.

Japan is facing new challenges and new problems and issues today. These focus on the environment and energy depletion, which relate to environmental ethics, high electronics, and biotechnology. These in turn are related to quality of human life and bioethics. The Japanese citizenry is not well prepared to solve these new problems and issues or to make responsible decisions. Nor are students who complete schooling and graduate from upper-secondary school in Japan. There is new opportunity to seek out answers to these problems and issues, as well as to discuss more about how issues and problems associated with environmental education and technology education could be linked to science education. All of these considerations continue to define and emphasize the STS reform efforts. The philosophical features of the STS approach are becoming a new paradigm shift for science education in Japan to meet urgent needs. As a consequence, the STS approach in the Japanese science education community is "poised for a forward leap" as observed by Yager at the close of his July 1993 visit in Japan.

Plans for the Future

Before making specific plans for STS as reform for Japanese science education, it is important to note the unique features of Japanese educational system. It is likely that Japanese society has not yet matured to develop the grassroots movement because the Ministry of Education assumes the burden of the curriculum implementation based on a national Course of Study. This is the so-called top-down system of educational administration.

However, if STS characterizes the reform, a new paradigm for science education in Japanese society will occur. Several considerations must be addressed if this does occur. First of all there is the short-term strategy, which should be promoted. Such strategies include goals that are easy to realize and whose effects will be easy to evaluate. Such strategies include developing STS modules based on the new goals as mentioned earlier. This should include pre-service training programs which in turn should assist with moves to STS at the college level. For example, some college science educators who have interest in and understanding of STS must conduct preservice training programs focusing on STS for prospective science teachers for both the elementary and secondary schools. They can offer STS instruction for the prospective science teachers and include the research rationale for it for undergraduate and graduate theses. As a consequence, STS teachers will increase in number and exert greater influence in schools. They will be involved with research and they will experience learning within the new STS context.

Second, STS's systematic approach can become a long-term strategy. We at the University of Tsukuba, a lead university, are trying to conduct the in- and preservice training programs for STS instruction as a major reform for science education: (1) seeking to compare strategies and to cooperate with STS advocates in other countries, (2) comparing STS classes and non-STS classes in terms of student learning, and (3) synthesizing the philosophical and practical points of view and to synthesize a new meaning for the Japanese situation. This role of the lead university and this system for a study plan for Japanese STS are summarized in Figure 23.1. Lead universities in other countries will conduct the research in and for their respective nations. Such trials and experiments around the world will help STS grow everywhere and thereby assure a general paradigm shift.

As shown in Figure 23.2, both Lead Teachers and lead universities will cooperate as Japanese STS emerges with a unique philosophical base and a focus on practical studies. The STS movement will be offered and evaluated as real reform in Japan. Eventually, the STS approach should be adopted by the Ministry of Education in Japan as a national reform effort. The Ministry officials are already engaged in such discussions at the highest levels. We hope that government endorsement and encouragement occurs in keeping with the philosophical and practical points of view arising from both short- and long-term strategies. We hope that the STS movement in Japan can become a grassroots model for promoting the STS movement in the near future.

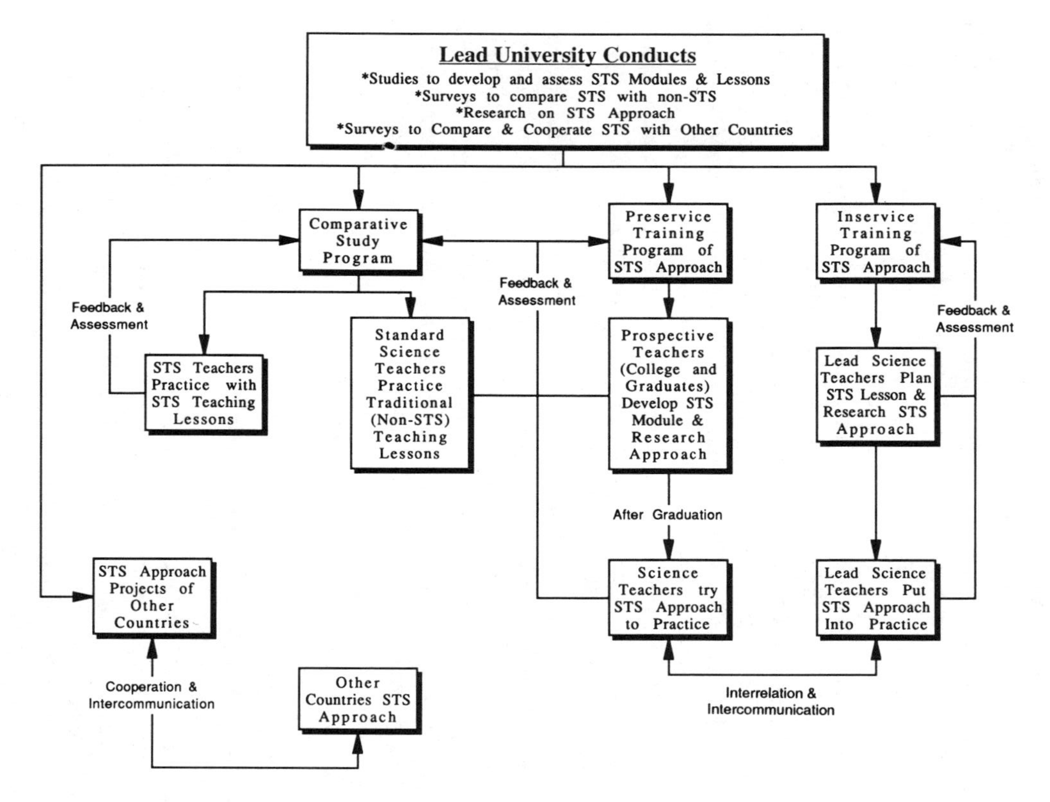

FIGURE 23.1 Study Plan for Japanese STS

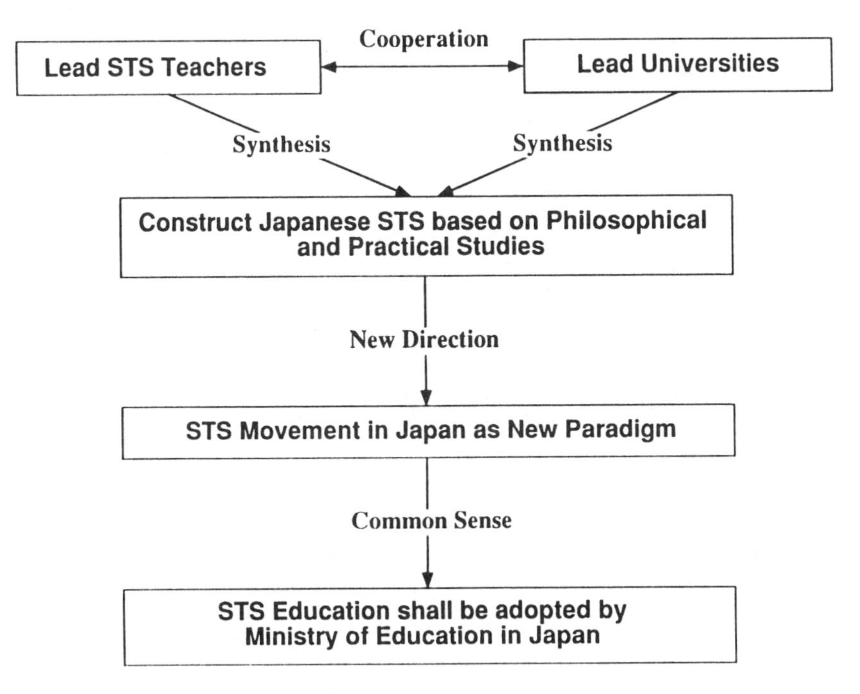

FIGURE 23.2 Japanese STS Movement

REFERENCES

Kellerman, L., and Varrella, G. (1993, January). *Criteria for change from the traditional to the STS approach: An Iowa perspective.* Paper presented at the annual meeting of the National Association of Science, Technology and Society, Arlington, Virginia.

Kumano, Y. (1991). STS approach and environmental education—present tendency of science education in the U.S.A. *Journal of Science Education in Japan, 15*(2), 68–74. (Japanese.)

Kurioka, S., and Nogami, T. (1992). Philosophy of SATIS project and its teaching methodologies. *Bulletin of Society of Japan Science Teaching, 33*(2), 17–26. (Japanese.)

Ministry of Education. (1991). *Elementary teachers' guide book for environment education.* The Ministry of Education. Tokyo. (Japanese.)

Ministry of Education. (1993). *Secondary teachers' guide book for environment education.* The Ministry of Education. Tokyo. (Japanese.)

Morohashi, S. (1983). SISCON project: College education in the U.K. *Journal of Science Education in Japan, 7*(3), 113–120. (Japanese.)

Nagasu, N. (1987). The recent trend of science education in the U.S.A. In *Science and Mathematics Education in the U.S.A. Report of Cooperative Study on Education in Japan and the U.S.A.: Science and Mathematics Education Team.* National Institute of Educational Research, 10–77. (Japanese.)

Nagasu, N. (ed.) (1991). *Rationale of STS education and development of STS program in Japanese secondary biology education.* Report by Funding Minister of Education: Grant-in-Aide (A) No. 02301107 (Principal Investigator Umeno Kunio). (Japanese.)

Nagasu, N. (1993). What is STS approach: Historical and practical background. *Bulletin of Society of Japan Science Teaching, 33*(3), 79–89.

Nagasu, N., Yoshihiko, T., Kainamu, K., Tanzawa, T., and Antonio, B. E. (1992, May). *STS approach: Could it be a new challenge to science education in Japan?* Paper presented at XVII Pacific Science Congress: Science/Technology/Society: A New Context for Science, Honolulu.

Ogawa, M. (1989). *Science and technology education as civic education on a scientific technological society.* Casio Science Foundation Report, Ibaraki University. (Japanese.)

Ohsu, R. (1991). Basic study on STS education—STS education and philosophy of science. *Bulletin of Society of Japan Science Teaching, 31*(3), 20–23. (Japanese.)

Suzuki, Z., and Harada, T. (1990). Study of the interrelation of environmental education and the STS education. *Bulletin of Osaka University of Education*, 85–94. (Japanese.)

Takeuchi, Y., and Nakajima, H. (1989). *Teaching and learning about science by Ziman, J* (translated into Japanese) Tokyo: Sangyo Toshio.

Tanzawa, T. (1992). Japanese science teachers' perception of science and technology related global problems and STS. *Journal of Science Education in Japan, 16*(3), 115–125.

Umeno, K. (ed.) (1992). *Development of teacher training and secondary biology education based on STS education.* Report by Funding Ministry of Education: Grant-in-Aide (A) No. 02301107 (Principal Investigator Umeno Kunio). (Japanese.)

Watanabe, S., and Ikeda, H. (1992). STS education in the U.K. *Heredity, 46*(12), 54–59. (Japanese.)

CHAPTER 24

STS INITIATIVES IN AUSTRALIA

Geoffrey Giddings

NATIONAL CURRICULUM FRAMEWORK

In April 1989, a meeting of the Australian Education Council (AEC), which comprises Commonwealth, State, and Territory Ministers of Education, ratified a historic agreement that describes ten Common and Agreed National Goals for Schooling in Australia. Following this agreement the AEC agreed to support the development of national curriculum statements and profiles for eight areas of learning. One of these areas of learning is science. Thus, the National Science Statement (DEET, 1992) is one part of a broad program of national collaboration in education—a difficult path for the States to follow, due to the fact that traditionally the individual States and Territories have controlled most aspects of educational delivery since the federation of Australian States.

In this Science Statement it is acknowledged that science education has an important role in achieving all of the Common and Agreed National Goals, and in particular: "Science education will develop an understanding of the role of science and technology in society, together with scientific and technological skills" (DEET, 1992, p. 2). Hence the identification of significant goals in the STS area are very much part of current thinking in science education reform in Australia.

SOME RECENT HISTORY

During the early and mid-1980s, Australia participated in the Second International Science Study (SISS) conducted by the International Association for the Evaluation of Educational Achievement (IEA), which assessed science achievement across seventeen countries. The international coordinator for the study was an Australian, Dr. Malcolm Rosier of the Australian Council for Educational Research (ACER). Another Australian, Dr. John Keeves, former director of ACER, was the chairman of the international policy committee for the study. The IEA tested the achievement of children aged ten, fourteen, and seventeen in 10,000 schools throughout these countries and concluded that, in comparison to the other countries in the sample, Australia had slipped dramatically since the last tests were carried out in 1970 (Jacobson and Doran, 1988). In 1970 Australia ranked third in the seventeen-country sample; in the most recent testing, the nation slipped to tenth position (Rosier, 1988).

Unfortunately, science education at primary, secondary, and tertiary levels in Australia is beset with several serious problems in addition to these declining standards (Rosier, 1988). Evidence of the declining participation rates in science and technology disciplines at the secondary and tertiary levels is irrefutable. In the traditional secondary science courses such as physics and chemistry, the proportion of twelfth-year students taking such courses has declined nationally over the past ten years (Dekkers, de Laeter, and Malone, 1986). The situation is mirrored at the tertiary level: whereas total tertiary enrollments have almost doubled over the past decade, enrollments in the physical and geological sciences have not kept pace with this overall increase. The Commonwealth Schools Commission's *National Policy for the Education of Girls in Australian Schools* (Australian Federal Government, 1987) highlights the concern that, of all school subjects, perhaps the greatest inequity between the sexes in enrollment, achievement, and attitude occurs for the physical sciences (Fraser & Giddings, 1987). The time is ripe in science education to redress these low enrollment numbers and disappointing achievement levels for all students. Such an exercise is more likely to become a reality if there are strong and clear national guidelines for both science and technology education.

Science and Technology Education in Australia

Setting and achieving a national agenda for science and technology in education is a top priority for Australia's policymakers in this, the last six years of the twentieth century. Teachers and schools are constantly being told through the media and other sources that scientific discovery should be related to technology and that as the resulting effects of the technology on society has such an important impact on peoples' lives, that schools should be including STS courses in the school curriculum. Common criticisms of Australian science curricula usually assert that students are being taught a curriculum that is essentially irrelevant, too formalized, and often disconnected from the lives most people live.

Hence, much interest has been generated in STS courses in Australia. Cross (1990) alerts us to the range of meanings attached to the STS label that have become apparent at international conferences and through a variety of publications and programs in many different countries. The STS acronym is seen by Australians as the banner given to a genre of approaches to the teaching of science. Indeed, it is clear that there are several identifiable agendas underpinning the different values inherent in the various STS approaches. For instance, there are clearly science teachers in Australia who are greatly concerned about the problem of ensuring that students connect science with the social, technological, and political forces that shape the field, and that STS approaches appear to offer an excellent vehicle for achieving these ends.

Before discussing some of the initiatives on the Australian scene, it should be stated that Australian educators, science and others, have been carefully formulating and evaluating their views in this area, in order to achieve the goals they wish to achieve, in the most effective manner possible. The Australian approach to STS could therefore be described as very positive, but questioning.

Science teachers in this country are alert to the need for a questioning approach to STS in order to ensure compatibility with their own educational objectives. The role and nature of the term "technology" in the STS equation is a crucial one for Australian educators. On the one hand, much of the STS material appears to promote teaching for and about technology in a language that expresses an optimistic viewpoint about technology. Such an approach has been promoted in this country by the national government, although there have been words of caution. For instance, in 1982 Barry Jones, the national Minister responsible for Science and Technology at the time, claimed that "every technological change has an equal capacity for the enhancement or degradation of the quality of life, depending on how it was used . . ." (p. 231). On the other hand, a more socially responsible approach as advocated by Cross (1990) may be an evaluative one that adopts a neutral stance toward technology.

Certainly there has been a changing view of technology. There has been a shift from considering technology as merely the application or use of science, to one that portrays technology as involving problem-solving, drawing on knowledge and skills from several disciplines and hence developing the capacity to apply knowledge for some human purpose (Kings, 1990). Technology curricula in this country have tended to grow out of science (e.g., STS), industrial arts (Craft-Design-Technology [CDT]), or be developed independently.

No single approach dominates at the present moment—the different States are busy developing and designing their curricula along statewide guidelines, given that the nation is only just developing a single national imperative on this issue. That is not to say, however, that this situation will remain static. Fensham (1988), when commenting on the variety of STS approaches to science teaching, argues that where the science content is highly prescribed (as is the case in many Australian states), an STS approach of an "add-on" variety without much depth of treatment is likely. In those cases where the science content is not so highly prescribed there is the scope to formulate more substantive STS treatments. These comments provide a reasonably accurate summary of the situation across the States in Australia in the early 1990s. With the strong moves toward core national curriculum statements from the national government, it is likely that STS approaches will move from being a major "add-on" to the status of a main recurring theme.

STS has also been important for the various Australian states to carefully consider the amount of time that can be devoted to technology and/or the technology component of science education and whether teachers have the full range of skills required to teach this area. Certainly the traditional science discipline background of Australian science teachers does not equip them adequately to handle many of the social and controversial aspects of their teaching.

EXAMPLES OF SCIENCE, TECHNOLOGY, AND SOCIETY CURRICULUM REFORM ACROSS THE STATES

This section outlines some of the directions and the accompanying rationale for STS and related approaches in three Australian states—Victoria, New South Wales, and Western Australia. Kings (1990) reports that in Victoria there have been at least two relevant concurrent developments. The Blackburn Report (Ministerial Paper, 1985) on postcompulsory schooling (grades 11 and 12) and the *Curriculum Frameworks P–12 Report* (Education Department of Victoria, 1985). Both had profound implications for science and technology curriculum reform. Based on the recommendation of the Blackburn Report, science courses in this state have been structured to include STS-type themes. Based on the Curriculum Frameworks document, the purposes for science as stated in the curriculum guidelines include the following—science as knowledge, science as technology, science and society, science and personal development. Within the boundaries of these purposes, teachers are developing programs that give greater emphasis to what is termed the "social context of science and technology." This phrase or theme would seem to encapsulate the main common denominator of STS Australian-style.

It should be pointed out, however, that within the Australian context, all curricula (except at grade 11 and grade 12) are currently broadly based and individual schools decide on the details of what is to be covered in any given curriculum. In Victoria again, new curriculum initiatives in information technology studies have evolved quite independently of other studies (including science) and appear to conform to acceptable definitions of technology. The argument that technology education has little, if anything, to do with science has a degree of support in Victoria. Its proponents are developing technology education in such a way as to have its own curriculum, its own knowledge and skills, its own equipment, its own space, and its own new breed of teachers (Fensham, 1990).

On the other hand, many new science courses in Victorian schools have included units of study that clearly fall under the STS approach, but for various reasons are not specifically designated as such. These new courses include units with such titles as: Transforming Raw Materials and Meeting Human Needs, Transforming and Reshaping Products and Meeting Society's Needs, Extending our Capabilities and Creating New Hazards, and Science and the Development of World Views.

It should be mentioned that a number of overseas curriculum innovations, such as the British Science and Technology in Society (SATIS) materials, are fairly well known in Australia (Holman, 1986) and extensively used as "add-on" applications for traditional science courses. As Fensham (1990) points out, these add-on approaches are easier for teachers and do not disturb the overall course structure too much. It also means that teachers can now control how much technology they do, and allows the needs and constraints in their own particular school, their own background knowledge, the availability of relevant equipment and materials, to be highly influential in the nature of the course they construct and teach.

Other initiatives that have been important factors in raising the profile of the STS connection have been the compacts and partnerships developed between industry (private companies) and schools and colleges. Some outstanding examples of this collaborative approach to science and technology education reform have their genesis in Victoria (NBEET, 1993), but many are truly national. These collaborative arrangements include such initiatives as: scholarships for exceptional children, site visit programs, the sponsorship of national videos and other media material, and the organization of national Science Summer Schools and programs for the gifted and talented. Other companies have forged specific School-Industry partnerships aimed at creating strong links between a particular school and an industry partner. One particularly successful program of collaboration at Wesley College (Melbourne) involves four components: a scientist-in-residence program; a staff development focus; the supply of additional equipment and materials; and provides work experience for students and teachers.

In the Western Australian state context, one of the outcomes of the desire to implement technology education in schools was the decision in 1987 by the state Ministry of Education to designate and fund six pilot secondary schools to be designated technology high schools. The way each school proposed to integrate technology into its school depended on its definition of technology as well as the knowledge and skills of the teachers in the school. This section is a summary of this particular initiative from the perspective of one of these schools, derived from an early review of the innovation (Treagust and Mather, 1990).

The desired outcomes for technology education chosen at the particular pilot school after much debate, were basically those identified by the Commission on Technology Education for the State of New Jersey (Fricke, 1987). Using these as a starting point, the school developed and implemented an across-the-school technology education curriculum model. This curriculum model integrated technology objectives across *all* subjects in the compulsory grades eight through ten.

Each subject area rewrote the objective in its units incorporating the four technology areas selected—technological literacy, technological awareness, technological capability and information technology. Specifically, the science curriculum for grades eight through ten at the pilot school now has a range of technology objectives integrated within it. Interestingly, the science staff decided that not all units would have technology objectives integrated within them. Their preference was to include them initially in all the "core" science subjects that all students would study. Student reactions to the technology aspect of their curriculum was whole-heartedly positive. The manufacture of clothes in the home economics unit Technology and Fashion illustrated an increased understanding of the role of technology in all aspects of clothes making and manufacture. In the science area students met their community- and commercial-oriented projects with interest, commitment, and enthusiasm. One such course in robotics, utilizing Lego-Technic sets, involved prototype building and ultimately an exploration of the marketing and commercial potential of products of the course.

This project is clearly an ongoing curriculum initiative and one that "has the potential to be used by other secondary schools throughout Australia . . . [but] requiring a sound commitment among the teaching staff ensuring they are implementing the desired technology education at the school" (Treagust and Mather, 1990, p. 59).

The state of New South Wales has many good examples of what we may call a more conventional STS approach to science teaching. One such example is the Science, Technology, and Society Curriculum designed for the Department of Education (NorthWest Region) by Paterson, Boggs and Patterson (1988). The course has been written to meet a perceived gap in observed in the senior school curriculum offerings in science. The course has been specifically written to link the three strands of science, technology, and society together. The course has been designed as a set of independent modules.

Teachers are free to choose any combination to satisfy the number of units required as set out below:

Grade 11	1 unit—4 modules	Grade 12	1 unit—3 modules
	2 unit—8 modules		2 unit—6 modules

Following traditional STS-style objectives, the main aim of the course is to provide students with an understanding of the impact of science and technology on the individual's quality of life, within a changing Australian society. Specifically, the course aims to:

1. Provide students with a knowledge and understanding of the interrelationships between science, technology, and society.
2. Provide students with a variety of practical experiences involving experimental work, information processing (including computer operation), modeling, games, and simulations.
3. Enhance the problem-solving, reasoning, and communication skills of the students so that they may develop competent and confident self-images.
4. Promote in students a positive attitude toward Australia's future, and a capacity to critically evaluate technological developments in terms of their impact on Australian society and the Australian environment.

The course consists of fifteen modules. Within each module the set of content ideas are not meant to be of equal time duration or prescriptive. The modules consist of: Introductory Core Unit-Technology, Science, and Society; Legal Drug Use; Living with Natural Materials; Clean Living; Food and Agriculture; Moving People; Information Technology; Energy and the Future; Science and Technology for the Consumer; Medicines and Diseases; Environmental Hazards to the Human Body; Modern Materials Technology; The Exploration and the Colonisation of Space; Electronics; and Biotechnology and the Future of Humans. All the evidence at this point in time suggests that this type of course is popular with both teachers and pupils, although clearly putting additional strain on the mental, physical, and material resources of the schools.

Summary

Science and technology continues to pervade and interact with Australia's economic, cultural, and social life as they do in most other nations. It is clearly our responsibility as teachers, educators, and decision-makers to present science in its real-life societal context. Through an STS orientation to science education, all students can be increasingly able to take part confidently in public debate and decision-making about science and public science policy. Students must be exposed to the skills and knowledge about science and its related technologies, about the successes and failures of these endeavors, and about the criteria one can apply to evaluate the ideas and products of this union. All citizens need to be party to, and in charge of, the decision-making processes that are shaping their lives. For Australian science teachers there is little debate about the need for the place of technology in their curriculum—what is not as clear is whether technology is sufficiently interdisciplinary to warrant an across-the-school approach to its teaching, and whether the teachers are trained appropriately in the skills to teach it. If the STS movement in Australia and overseas can claim a role in fostering an appreciation of the part science plays in shaping our cultural and intellectual heritage, then this will better enable our students to be participating members of both the global and Australian communities.

REFERENCES

Australian Federal Government. (1987). *Higher education: A policy discussion paper.* Canberra: Australian Government Publishing Service.

Cross, R. T. (1990). Science, technology and society: Social responsibility versus technological imperatives. *Australian Science Teachers Journal, 36*(3), 33–38.

Dekkers, J., de Laeter, J., and Malone, J. (1986). *Upper secondary school science and mathematics enrollment patterns in Australia: 1970–1985.* Perth: Western Australian Institute of Technology.

Department of Employment, Education, and Training (DEET). (1992). *Science for Australian schools: Interim statement.* Canberra: Australian Government Publishing Service.

Education Department of Victoria. (1985). *Curriculum frameworks P–12: An introduction.* Melbourne: Curriculum Branch.

Fensham, P. J. (1988). Approaches to the teaching of STS in science education. *International Journal of Science Education, 10*(4), 346–356.

Fensham, P. J. (1990). What will science education do about technology? *Australian Science Teachers Journal, 36*(3), 8–21.

Fraser, B., and Giddings, G. (eds.). (1987). *Gender issues in science education.* Perth: Curtin University of Technology.

Fricke, G. L. (1987). *Report to the commission on technology education for the state of New York: Technology education.* Learning how to learn in a technological world. Trenton, N.J.: New Jersey Technology Education Project.

Holman, J. (ed.). (1987). Special issue on science and technology in society. *International Journal of Science Education, 10*(4).

Jacobson, W., and Doran, R. (1988). *Science achievement in the United States and sixteen countries.* New York: Teachers College Press, Columbia University.

Jones, B. J. (1982). *Sleepers awake! Technology and the future of work.* Melbourne: Oxford University Press.

Kings, C. (1990). Developments in science and technology education: Some cross-cultural comparisons. *Australian Science Teachers Journal, 36*(3), 39–45.

Ministerial Paper. (1985). *Ministerial review of post-compulsory schooling.* (The Blackburn Report.) Melbourne: Victorian Government Printer.

National Board of Employment, Education and Training (NBEET). (1993). *Issues in science and technology education: A survey of factors which lead to underachievement.* Commissioned Report No. 22. Canberra: Australian Government Printing Office.

Paterson, M., Boggs, D., and Patterson, G. (1988). Science, technology and society. Department of Education, North West Education Region, Sydney, NSW.

Rosier, M. (April 1988). International comparisons of science achievement. *ACER Newsletter*, *62*, 1–3.

Treagust, D. F., and Mather, S. H. (1990). One school's approach to technology education: Integration across the curriculum. *Australian Science Teachers Journal, 36*(3), 50–60.

CHAPTER 25

STS IN DEVELOPING COUNTRIES IN THE PACIFIC

Jack Holbrook
John Craven
Martha Lutz

Science, whether studied under the name of nature studies, environmental studies, health, or hygiene studies, is required in schools throughout the world. Regardless of course title, science has become an integral part of the school curriculum at both the primary and junior secondary levels. Globally, school systems are recognizing the pervasive role of science in society. In contrast, many of these school systems have yet to identify the significant role and impact technology has on individuals and society. As a result, technology is rarely offered as more than a technical or vocational course at the secondary level. To counter this deficit, the STS movement offers an approach wherein science and technology curricula are linked and redefined along more relevant lines. Throughout the STS approach, particular attention is given to the provision of quality science/technology education while emphasizing the students' role in the learning process. This is reflected in one of the basic tenets of the STS approach—that the learning of science content out of context is neither useful nor interesting. Two questions thus become apparent. How relevant is the STS model in the curriculum of developing countries? And what steps are being taken for its introduction?

Currently there are signs that a rethinking of education is taking place and that striving for "relevancy for all" is being considered as a fundamental goal. As a result, science education at the primary, middle, and secondary level is being geared much more to indigenous technologies as well as the development of skills beyond those of factual recall or the acquisition of abstract concepts.

This ideological change is apparent in the Botswana junior secondary-science curriculum (1985). In a section on how to keep healthy, students are required to:

1. plan a balanced meal, know how to keep food free form contamination;
2. know the importance of good dental care, list essential childhood vaccinations;
3. be aware of the importance of regular exercise; and
4. discuss adverse effects of smoking, drugs, and alcohol.

The intent of these changes has been that science be seen as not only useful but requisite for living in society. Science and technology, seen in this light, then makes the integration of conceptual understandings with daily experiences both relevant and necessary.

Increasing attention toward focusing on the relevancy of curriculum is occurring in other developing countries as well. Many attempts have been made to integrate science education with local concerns. For example, the junior secondary-science course in Botswana, a dry country, emphasizes water conservation. Zambia has developed an agricultural science curriculum relevant to the needs of the country. The science in Ghanaian Society Project tries to promote local, indigenous technology, and incorporate this into the science curriculum.

The science for Ghanaian society project (Yakabu, 1990) advocates concerted efforts to incorporate into the science curriculum student projects, debates, role-playing exercises, and experimental investigations. This project deliberately attempts to be a program that incorporates the STS framework and, thereby, guides students to appreciate the links between themselves, science, and society.

Another program whose intent was to incorporate the STS instructional approach is illustrated by a certificate level (grades 10–11) chemistry course in Hong Kong (1991). The curriculum guide has been developed to rethink the "fundamentals first" approach. This guideline suggests a teaching approach that leads from the product to the concept. For example, an introduction based on a familiarity with detergents leads to source of detergents, properties of detergents, and issues related to the use of detergents. The section on fossil fuels begins with crude petroleum and progresses from petroleum distillates to the demand for these distillates. In progressive sequence the introduction of homologous series, burning of fuels and fire fighting, environmental impact issues, and, finally, alternative energy sources follows. Suggested activities include:

1. Experimentally testing and comparing soaps and detergents;
2. Projects centering on air pollution issues; and
3. Debates on the desirability of using fossil fuels or alternative energy sources.

There have been obstacles, however, to the implementation of the STS approach in such countries. It is important to note that many developing countries have inadequate laboratory facilities, science equipment, and supplies (Charakupa, 1991; Van den Berg and Lunetta, 1984; Urevbu, 1984). Additionally, there are budgetary restrictions that contribute toward decreasing state expenditures per pupil in nonsalary areas such as textbooks.

In addition to budgetary restrictions, there are also philosophical obstacles to the implementation of the STS approach in the classroom. Science programs used in developing countries are often directly incorporated from Western nations' science programs (Ingle and Turner, 1981; Ogawa, 1986; Williams, 1979) with little regard to local customs or beliefs. Traditionally, the education system within many developing countries still remains tied to their past history, and as a result, the local people view science in terms of their local culture (Kahn, 1990). The local interpretation of what science is often leads to the problem that Western terms or concepts have no traditional equivalent. Indeed, these terms or concepts are often opposed to local cultural values (Kay, 1975).

There are some encouraging, albeit limited, signs that local adaptation of STS is occurring. The Tongan and Tuvalu syllabi reflects such local adaptation of STS. Tuvalu is endeavoring to promote public awareness about AIDS and environmental awareness. However, there is little integration into the educational system. Western Samoa has printed locally produced study units but has no formal textbook either written or adapted for Western Samoan schools. Other South Pacific countries have reorganized the curriculum materials so as to improve student interest and to include relevance to local needs as well as to address current world issues such as the environment. In the Solomon Islands, for example, the main thrust of the science curriculum is to develop (in all students) an understanding of science and improve skills that will be applicable to their lives whether they live in the village, town, or city. Currently, there are two types of secondary schools in the Soloman Islands. One is designed for more academic career courses while the other is intended for vocational, STS-oriented studies. The schools moving toward the STS approach seek to develop students who are capable of contributing positively to their local area once they have finished their formal education. As a result of the reoriented goals, the Solomon Island schools have produced a considerable range of locally adapted science materials. In a final example, the Curriculum Development Unit of Papua, New Guinea, has adapted an Australian science textbook for use in secondary high schools. The adaptation includes a strong Papuan flavor by emphasizing local technology and using photographs of Papuan scenes. Currently, units are being rewritten for a reformed education system with stronger emphasis on relating traditional stories or practices to the "scientific approach." While these changes are encouraging, conflicts between educational systems, individuals, and society must be reduced in order the accelerate the reform efforts and achievement.

But cultural and conceptual conflicts are not unique to developing countries. Similar conceptualization problems arising between learning in society and learning in school have been an areas of considerable research in Western countries. This research was the original impetus that led to a new look at the constructivist approach to learning (Driver, 1993). In this approach, the teaching of science in particular is based on previous learning experiences and does not treat students as empty vessels that need only to be filled with abstract conceptual ideas. Development of STS programs is more critical in developing countries: conflicts that potentially could arise from differences between the school and the home are recognized and care should be taken that two different sets of values and attitudes do not arise (Kay, 1975). Aside from cultural differences, two major factors inhibiting STS-style teaching in many countries around the world, namely: (1) the dominance of the textbook, and (2) the perceived need for formal, external examinations.

First, the textbook, often written in a "fundamentals-of-science-first approach," places scientific facts and principles as the major component. The pattern of such textbooks hardly encourages alternative or "science in the society" approaches. All too often, applications of scientific principles translate to being little more than a few pictures showing its use. Activities such as debates on issues, role playing, or even investigations that call for recognition of the problem, followed by planning the investigation, manipulating the variables, and then interpreting the results are hardly areas that can be simply incorporated into a reference style textbook.

And second, to date, examination boards have refused to change the style of their examinations because they do not wish to risk loss of reliability in exchange for modest growth in validity and because they contend that the pencil-and-paper tests ensure objectivity, are cost-effective, and provide uniformity. In short, many central examination boards do not trust the ability and judgment of teachers. Many of these examination boards feel that their measurement practices worked well in the past so why change now? For instance, high scores in an international study of science achievement are used to confirm achievement of Hong Kong students (Holbrook, 1990). However, this conclusion is misleading in that the test measured achievement of students committed to science and who have succeeded in science in the past. It did not measure science achievement of all students.

The emphasis and importance of achievement is strongly rooted within the educational philosophy of many developing nations. Oftentimes, because of poor economic conditions, acquiring good science education is perceived as a potential means to escape poverty. Science education is seen primarily as a way to become a scientist, or to get a government job, or perhaps to travel to a developed nation for further advanced training. The traditional background of education in developing countries has been a competitive system with limited opportunities for advancement for those unable to pass fact recall examinations. Because of this view, parents of students in developing nations are generally unwilling to accept any innovations in science education that do not obviously lead to privileged jobs for their children. The idea of scientific literacy, or science for all, is unwelcome because it may deflect educational goals from promoting their children to a higher standard of living.

An example of this kind of resistant force can be found in New Guinea. In the early 1980s Vulliamy (1988) attempted to diversify the Papua, New Guinea, secondary school science curriculum to make it more relevant to rural development. He intended that the grades nine and ten science curricula would be adapted to include projects that students would complete while living at school-outstations (thus giving the curriculum an STS approach). The success of this project depended on appropriately trained and enthusiastic teachers. His curriculum was piloted for a number of years but has since been largely abandoned due to both limited funding and to skepticism on the part of some participating teachers. Generally, teachers did not perceive the project as preparing the students for an academic future, and so they did not support the concept of the project. Also, parents demanded that secondary schools prepare their children so that they could be eligible for tertiary education. They perceived the programs as being second-rate education for their children (Turner, 1990). Parents perceive schooling for their children as a means of escape to a more privileged sector of the work force (Bacchus, 1984). This is because tertiary education often brings a government job.

This pattern is now slowly changing. The "Education for All" movement (WCEFA, 1990), initiated at a world conference in Jomtien, Thailand, is attempting to mobilize support for primary education for all children worldwide before the year 2000. In addition, more space is being made available at the secondary and tertiary levels. Further, adult education programs are being developed. To facilitate such changes, programs that demonstrate to parents, students, and local educators that the STS way of teaching science will not strain their limited educational budgets. Community members must be shown that along with improved educational outcomes, all students will have a greater potential for higher achievement that in turn could lead them to a higher standard of living. These efforts are prerequisites to sustaining educational reform. The increasing influence and impact of technology on society may provide the impetus for change. Globally, there are similar changes occurring in both science education and scientific communities.

Yager (1992), in his article on science in the Pacific Region, discusses how science education is currently undergoing just such a paradigm shift. The nature of this shift is a focus on technology, as opposed to "pure" science. Technology is relevant to modern life, is intrinsically interesting (particularly to young people!), and has the potential to provide good jobs and raise the standard of living. However, it is not necessary to copy exactly the curricula of other nations in order to achieve successful educational outcomes. For example, although it is becoming increasingly common for schools in the United States to have computer facilities, and to use computer simulations as part of science training, it is possible to have an excellent science education program that makes no use at all of computers. If one cannot afford computers, it is better to make do without them than to abandon all attempts to improve science education for lack of some resources.

There are two concepts that must be kept in mind when redesigning science curricula in developing nations. One is the emerging evidence that suggests that science learning is improved when it is centered around personal relevance. Mounting evidence suggests that students learn science best when concepts are presented in context and related to their daily lives. Knowledge that is presented in isolation is not learned, or, at best, not well retained. The other factor is the need for actual experiences that could be hampered by limited funds available for science education.

These two elements can be dealt with simultaneously. Research of STS teaching practices is beginning to yield encouraging results in the United States and England. Evidence mounts that STS not only improves learning in terms of retention of content, but also attitude toward science improves. Evidence indicates that STS teaching competes successfully with traditional teaching and may eventually prove far more effective. By definition, STS is personally relevant; in practice STS provides science learning in familiar contexts. The perception that a science laboratory must be provided with expensive equipment in order to supply quality science education is obsolete. In addition, many STS investigations can be carried out with a minimum of special equipment. Excellent science has been and continues to be practiced with a minimum of costly equipment in many regions. Quite often, these investigations rely on the inclusion of indigenous resources (George, 1993).

One possible scenario is the use of a homemade Berlese Funnel to investigate soil fauna. The results of such a study could be used to draw conclusions about issues of health or agriculture. A functional Berlese Funnel can easily be made from a large coffee can (or the equivalent), a scrap of metal screen, a plastic cup, a baby food jar (or the equivalent), and a single light bulb. The Funnel can be used as a takeoff point for discussions of heat, light, dehydration, ecology, and animal diversity. The soil fauna collected via use of the Funnel could be identified, put in a collection, studied as examples of local animal diversity, and even reported to an entomological journal. Think how exciting it would be for students to see their STS project written up and published in a scientific journal.

The idea of making and using a Berlese Funnel is only one of a nearly infinite variety of possibilities that are open to STS teachers. Teachers and students could explore basic principles of physics by making a "telephone" from paper cups and a string—something almost every child in the United

States used to know how to do—before expensive equipment began to threaten creative thinking. Science educators need to lead the way in a rediscovery of the tremendous potential of simple materials that can be used to investigate science.

Although it is almost certainly beyond the scope of science educators to improve science education in developing countries by improving the economy, it is reasonable to believe that science educators could help improve the economy of developing countries by improving the quality of science education. And the best hope for that is to encourage a paradigm shift: science education should center around STS-type investigations of local issues. These investigations should freely include the unique and inexpensive indigenous resources. Catalogues of indigenous resources should also include local art forms and folk tales. George (1993) describes how a teacher in the West Indies successfully incorporated calypso and masquerade into science education and, thereby, provided personal relevance and a dimension of creativity for her students.

In conclusion, to overcome the economic barriers to improving the quality of science education in developing countries, science educators need to assist those countries in implementing STS teaching techniques. This should include: demonstrating the improved understanding and retention promoted by STS teaching; encouraging use of indigenous resources; discouraging a view that expensive equipment is required for reform; and promoting the view that science education can be beneficial for all, not just a privilege for a select few who intend to improve their standard of living. Thus far, intentions outdistance implementation. More evidence is needed to show that a worldwide "science for all" movement is underway. There are many factors preventing this. Teacher education needs to be more comprehensive to enable teachers to cope with new content and new skills. Inservice support for teachers needs to be strong if teachers are to "change" along the lines of the new developments. Research has repeatedly shown that teachers, left to their own devices, teach as they themselves were taught.

A new movement, initiated by the International Council of Associations for Science Education (ICASE) and UNESCO, is designed to enhance the implementation of scientific and technological literacy for all. This program intends to target basic education levels and adults through nontraditional education programs. Project 2000+ (1993) intends to promote scientific and technological literacy by considering six areas including:

1. The need for and nature of scientific and technological literacy;
2. Scientific and technological literacy needs for development;
3. The teaching and learning environment;
4. Teacher and leadership education;
5. Assessment and evaluation mechanisms; and
6. Informal and informal methodologies.

The plan calls for the creation of national task forces to initiate and oversee developments in scientific and technological literacy not only with respect to curriculum change, but to dissemination and implementation mechanisms as well. Assessment of scientific and technological literacy will also be an important issue to be tackled through research and developmental projects. The new movement recognizes the need to promote an STS teaching style worldwide and also recognizes that this can only be achieved if those involved (curriculum planners, government officers, community leaders, teachers, and teacher associations) can join together to ensure that implementation is supported and monitored. This is a monumental task, but an exciting challenge that ICASE, UNESCO, and its other partners in Project 2000+ are accepting.

References

Bacchus, M. K. (1984). Planning secondary education in a developing country: A study of Papua New Guinea. *Compare*, *14*(2), 189–203.

Botswana Ministry of Education. (1985). Integrated Science Syllabus. Junior Secondary Junior Certificate Syllabus.

Charakupa, R. (1991). Issues in the learning of science in Botswana secondary schools. *Science Education International*, *2*(3), 8–11.

Driver, R. (1993). A constructivist view of learning: Children's conceptions and the nature of science. In R. Yager (ed.), *The science, technology, society movement* (p. 103–112). Washington, D.C.: National Science Teachers Association.

George, J. M. (1993). Quality provision in science in an environment with limited commercial resources. *International Journal of Science Education*, *15*(1), 17–25.

Holbrook, J. B. (1990). *Science education in Hong Kong: Achievements and determinants*. Education paper 6, Faculty of Education, University of Hong Kong.

Hong Kong Curriculum Development Council. (1991). Syllabus for secondary schools. *Syllabus for Chemistry (secondary 4–5).*

Ingle, R. B., and Turner, A. D. (1981). Science curricula as cultural misfits. *European Journal of Science Education*, *3*(4), 357–371.

Kahn, M. (1990). Paradigm list: The importance of practical work in school science from a developing country perspective. *Studies in Science Education*, *18*, 127–136.

Kay, S. (1975). Curriculum innovations and traditional culture: A case history of Kenya. *Comparative Education*, *11*(3), 183–191.

Ogawa, M. (1986). Towards a new rationale of science education in a non-western society. *European Journal of Science Education*, *8*(2), 113–119.

Ogunniyi, M. B. (1988). Adapting western science to traditional African culture. *European Journal of Science Education, 10*(1), 1–9.

Project 2000+. (1993). *Report of an international forum,* Paris: United Nations Educational, Scientific, and Cultural Organization.

Turner, M. M. (1990). *Papua New Guinea: The challenge of independence.* Ringwood: Penguin.

UNESCO. (1986). *Summary report of science, technology and mathematics education worldwide.* Paris: United Nations Educational, Scientific, and Cultural Organization.

Urevbu, A. O. (1984). School science curriculum and innovation: An African perspective. *European Journal of Science Education, 6*(3), 217–225.

Van den Berg, E., and Lunetta, V. N. (1984). Science teacher diploma programs in Indonesia. *Science Education, 68*(2), 195–203.

Vulliamy, G. (1988). Adapting secondary school science for rural development: Some lessons from Papua New Guinea. *Compare, 18*(1), 79–91.

WCEFA. (1990). *Education for all: Purpose and context.* Monograph 1. Paris: United Nations Educational, Scientific, and Cultural Organization Press.

Williams, J. W. (1979). The implementation of curricula adapted from Scottish integrated science. In P. Tamir, et al. (eds). *Curriculum implementation and its relationship to curriculum development in science* (pp. 295–299). Jerusalem: Science Teaching Centre.

Yakabu, J. M. (1990). *Science in Ghanaian Society Project, Teacher's Guide.* Unpublished manuscript.

PART V

Supporting the STS Reform

Any reform is likely to dissipate without a support system. STS in the United States enjoys a relatively new professional society. A model for assisting students with moves to STS now exists with the Iowa Chautauqua Model. This model has been validated by the National Diffusion Network, which now means funding for its dissemination.

The new National Standards for Science Education have moved toward an STS focus since Koch wrote her chapter indicating a "lemon" version. The Standards now define science *content* to include eight dimensions, namely:

1. unifying concepts and processes;
2. science as inquiry;
3. physical science;
4. life science;
5. earth and space science;
6. science and technology;
7. science and societal challenges; and
8. history of science.

Nonetheless, the importance of evidence for STS successes continue to assist with a meaningful definition and the identification of needed changes.

Certainly establishing science and technology as an organizer for the entire curriculum elevates these fields to a central position, probably indicating the centrality of science and technology in today's society.

For realizing the full potential for STS, support is needed as well as acceptance that STS reform is dynamic and continues to grow—much like science itself does.

CHAPTER 26

NATIONAL ASSOCIATION FOR SCIENCE, TECHNOLOGY, AND SOCIETY

Stephen H. Cutcliffe

STS programs at U.S. universities are now celebrating their silver anniversaries, if one dates their beginnings with the establishment of formal STS programs at Cornell University and Pennsylvania State University in 1969 (Cutcliffe, 1989, 1990, 1993). A field of study can be said to have arrived with the establishment of professional societies and journals as outlets for the exchange of scholarly and educational pursuits. During the past two decades, a number of such organizations and publishing outlets have emerged, each of which contributes in important ways to promoting the central focus of STS, that is, the analysis and explication of science and technology as complex "social constructs" entailing cultural, political, economic, and general theoretical questions.

As the field of STS has developed, at least three different research and educational approaches can be identified: (1) Science, Technology, and Public Policy (STPP), (2) Science and Technology Studies, and (3) Science, Technology, and Society programs. The STPP approach is professionally oriented and generally has as its focus the analysis of large-scale sociotechnical interactions and their management, has a strong scientific and technical orientation, and stresses the need for and training in appropriate policy and management fields. *Technology in Society*, associated with the Society for Macroengineering, is an important STS journal with close ties to the technical community that typifies this approach. Science and Technology Studies in contrast involves more theoretical investigations into the social and cultural context of science and technology and their functioning as social processes. The Society for the Social Studies of Science (4S), founded in 1975, is the leading interdisciplinary professional society reflecting this approach. Its annual meetings and journal, *Science, Technology & Human Values*, tend to be focused on academic research with a sociological and policy orientation, although other perspectives are not ignored. Steve Fuller (1993) has characterized this explanatory and interpretive approach as "High Church" to distinguish it from the more problem-centered, social-activist oriented, or "Low Church" Science, Technology, and Society programs. This third STS approach, which in large part reflects the orientation of the National Association for Science, Technology, and Society, had its origins in late 1960s and early 1970s concerns regarding needed changes in undergraduate and, more recently by extension, in kindergarten through twelfth-grade education. It emphasizes general education and stresses scientific-technological literacy for intelligent responsible citizenship in today's scientific-technological society, as well as the contextual analysis of science and technology for its own sake. NASTS, a broad umbrellalike organization with a dominant educational focus that includes the kindergarten through twelfth-grade as well as collegiate levels, publishes its own periodic newsletter *STS Today* (formerly *NASTS News*), and is associated with the publication of the *Bulletin of Science, Technology & Society*. All of these groups, along with a host of others more disciplinary in orientation, are important to the stability and far-reaching impact of STS as a field of study; however, because of the educational theme of this volume, this essay will focus on NASTS and its role in supporting STS.

NASTS was established in 1988 to create an interdisciplinary forum, in particular through an annual conference and a journal, in which the wide range of individuals and groups expressing interest in and concern for the interrelationships among science, technology, and society could come together for the exchange of ideas and materials. NASTS chose as its logo a stylized umbrella in which various constituencies were identified as making up the specific panels. This logo suggests a level of interdisciplinarity for the field that went beyond an earlier more multidisciplinary approach that might have been captured metaphorically by a set of disciplinary focused groups, that is, history, philosophy, sociology of science, and technology, huddled together under the umbrella. The current NASTS logo eliminates the specific groups by name, but whether this suggests a progression to what Julie Thompson Klein in her book, *Interdisciplinarity: History, Theory, and Practice* (1990), has called "transdisciplinarity" or the creation of conceptual frameworks capable of influencing more than one discipline is not yet clear. However, it does suggest STS scholars and educators have begun to borrow, if not actually create anew, methodologies for common problem solving. Perhaps chief among them at this time is the conceptualization of science and technology as being "socially constructed" and only truly understandable when viewed in their full societal "context." By this STSers mean science and technology are inherently value-laden social processes that take place in specific contexts shaped by, and in turn shaping, human values as reflected in cultural, political, and economic, as well as scientific institutions. As such their understanding requires a holistic conceptualization of the complex interrelations.

In contrast to the more limited theoretical or political orientations of many other STS organizations, NASTS consciously attempts to encompass a diverse community of individuals with broad-ranging interests: kindergarten through twelfth-grade and postsecondary educators, policymakers, scientists and engineers, public interest advocates, science museum staff, religion professionals, print and broadcast media personnel, lay citizens, and representatives from the international community. Their interests are loosely grouped into five major theme areas: education and information; technology, industry, and work; ethics and values; health and biomedicine; and the environment. Of the five, education, and especially education at the kindergarten through twelfth-grade level, is clearly the strongest component of the membership, although all areas are certainly well represented. This broad-based effort is the mark that distinguishes NASTS and thereby separates it from most other STS organizations; however, it also presents some problems for the organization. In particular, some members find NASTS's inclusiveness too broad, incorporating as it does individuals at one end of the spectrum who are highly critical of the science and technology "as usual" approach, and those at the other end of the spectrum who are willing to advance little criticism and wish primarily to improve and extend public understanding and acceptance, albeit often through enhanced and innovative educational pedagogies of science and technology. This spectrum of views was, for example, vividly portrayed through a series of exchanges occasioned by Chellis Glendenning's (1990) essay, "Notes Toward a Neo-Luddite Manifesto," and several equally pointed member responses published in *NASTS News* (Henninger, 1991; Jarcho, 1990; Shamos, 1990; Williams, 1990). As a broad-based membership organization NASTS has sought to provide appropriate forums through its annual meeting and publications wherein one can explicate the complexities of science and technology by noting their positive and negative (both expected and unexpected) impacts, by analyzing how scientists and engineers go about their work, by scrutinizing the ways in which societal institutions contribute to the shape and development of science and technology, and by suggesting mechanisms for better control over the scientific and technological processes. It seeks to do so in a "constructively critical" way and should not be perceived as being "anti" science or technology. To do that would be counterproductive and uninstructive, and to think that NASTS, or STS more broadly, has such as its primary motive makes little sense and would be akin to viewing art critics as "antiart" (Winner, 1986, p. xi). NASTS instead seeks to provide a venue in which its members can meet, share concerns, and debate differing ideas on the nature of, and how best to handle, science and technology in today's society. To date it has been most successful in doing so in the educational arena.

NASTS convenes an annual meeting held early in the new year in the Washington, D.C. area. The meeting provides a venue for the broad membership to come together to hear leading plenary speakers address pressing issues of the day, to hear and make presentations on current research and teaching techniques, and to mingle informally with like-minded colleagues for the sharing of ideas. Among the invited speakers addressing the NASTS meeting in recent years have been the Reverend Jesse Jackson; Rosalyn S. Yallow, 1977 Nobel Prize winner in physiology of medicine for the development of radioimmunoassay; Lester R. Brown, founder and president of the Worldwatch Institute; F. James Rutherford, director of the American Association for the Advancement of Science (AAAS) Project 2061 science-education enhancement effort; Luther Williams, National Science Foundation (NSF) assistant director for Education and Human Resources; Jerry Brown, former governor of California and presidential candidate; and Donella Meadows, coauthor of *The Limits to Growth* and *Beyond the Limits*. The range of backgrounds, interests, and perspectives of even this short list of keynote speakers suggests the breadth and vitality of NASTS as an organization. However, perhaps the aspect of the annual meeting that attendees ultimately find most fruitful is the exchange of ideas regarding teaching techniques and materials in the numerous member-presented sessions devoted to such concerns. Here the primary focus has been on the kindergarten through twelfth-grade educational level, although collegiate efforts have by no means been ignored. At this level, STS is most widely found in the science education area, although efforts in technology education and the social sciences are also gaining headway.

The STS approach emphasizes teaching and learning science and technology in the context of human experience, in which real-world problems of interest to students provide the entry point, rather than concepts and processes to be memorized. Here the emphasis becomes one of process, empowerment of students, and responsible real-world decision-making. This approach to creating a scientifically and technologically literate citizenry for the next century is far more appropriate for most students than the traditional science education paradigm—*P*hysics, *C*hemistry, *B*iology, which Rustum Roy (1991), NASTS's primary founder, jokingly refers to as PCBs. STS, for which NASTS provides the most important educational forum nationally, thus tries to place an interdisciplinary, integrative real-world component at the center of general education, one that is important for all students, but especially so for those 90 to 95 percent who are unlikely to go on to careers in science or engineering.

As a professional institution NASTS has also begun to play an important role in shaping the direction of STS as a field of study and in terms of educational impact through the development, publication, and circulation of formal position papers. The first of these was the report of the Assessment and Evaluation Committee (Harkness, 1992), which argued for the expansion of assessment and evaluation standards and techniques to include technology and STS issues and themes in addition to the traditional science focus of current instruments. NASTS submitted this position paper to the National Science Education Standards project of the National Research Council (NRC) reflecting the association's perspective on the centrality of STS themes and approaches to science education. The work of the Science Education Standards project is still ongoing, so it is too early to tell the direct impact of such input, but clearly this is an important function of an education-oriented organization such as NASTS. A second-position paper formally accepted by the NASTS Board of Directors at its 1993 annual meeting was the report of the Position Papers Committee itself (Daugs, 1992), which laid out in formal fashion the principles that identify the STS approach. They include:

1. Connectedness: Emphasis on integration of parts into meaningful wholes.
2. Depth versus breadth: Personal relevance of a few topics is of greater value than encyclopedic coverage.
3. Collaboration: Cooperative, constructive learning is a must.
4. Individual empowerment: Everyone can contribute something to the whole. All contributions are valued.
5. Character development: Experiences should be structured to promote personal responsibility, integrity, honesty, critical thinking, and decision-making.
6. Technology and integration: Experience should involve appropriate technology processes and products. Technology should be viewed as a tool for expanding human potential.
7. Evaluation: No one set of values or procedures should be used to assess performance or evaluate individual performance.

NASTS, by developing such a set of principles, has thus established a framework to guide the pursuit of subsequent projects or positions on important educational issues.

A third important NASTS effort in support of STS is its publication program, both through its newsletter *STS Today* edited by Janice Koch of Hofstra University and the more formal *Bulletin of Science, Technology & Society*. The newsletter serves both to update the membership on association activities and as a forum for the discussion of STS issues in the schools, workplace, and the environment. It thus includes articles profiling academic STS programs, articles on educational issues such as the science education standards referred to earlier, and brief book and material reviews as well as announcements of appropriate events. NASTS is currently affiliated with the *Bulletin of Science, Technology & Society* (*BSTS*) edited by Rustum Roy and published by STS Press. *BSTS* was originally established in 1981 as a means of communication with the broad ranging STS community including not only educators at all levels, but also professionals in government and industry, working scientists and engineers, science and technology journalists, and the public. The focus of the journal as it has evolved has been primarily educational and pedagogical. Although *BSTS* includes original articles describing research or reflection on STS topics, they are generally of interest to the broader readership and of use in STS teaching. *BSTS* also includes both educational modules for direct use in the classroom and course outlines of interest to STS teachers. One particularly valuable feature of the *Bulletin* is that subscribers are automatically granted the right to reprint all original material published in the journal for classroom use, provided it is not resold for profit. *BSTS* also includes an extensive bibliographical listing of new books and articles in the STS field that may be of general interest to STS educators or of direct use in the classroom. This is one of the more valuable services of the *Bulletin*, for given the wide ranging nature of the STS literature, it is difficult for any one person to keep abreast of the current outpouring of material. NASTS through these two publications is thus able to reach out to communicate regularly with the broad STS community, including those many individuals not always able to come to the annual meetings.

In summary, NASTS is an organization with a fourfold mission:

1. to create a technologically and scientifically literate citizenry,
2. to integrate the humanities and social sciences with science and technology,
3. to help shape public policy that guides evolving technology,
4. to introduce and extend STS as an approach to teaching that starts with concrete everyday issues and experiences rather than abstract theory.

NASTS provides a needed venue for all those individuals who, whatever their particular viewpoint, share an interest in and concern for society's understanding and handling of science and technology in its societal context.

References

Cutcliffe, S. H. (1989). The emergence of STS as an academic field. In P. Durbin (ed.), *Research in Philosophy and Technology: Volume 9*, 287–301. Greenwich, Conn.: JAI Press.

Cutcliffe, S. H. (1990). The STS curriculum: What have we learned in twenty years? *Science, Technology & Human Values*, *15*(3), 360–372.

Cutcliffe, S. H. (1993). The warp and woof of science and technology studies in the United States. *Education*, *113*(3), 381–391, 352.

Daugs, D. R. (1992). The niche for STS. *NASTS News*, *5*(3), 3.

Fuller, S. (1993). *Philosophy, rhetoric, and the end of knowledge: The coming of science and technology studies*. Madison: University of Wisconsin Press.

Glendenning, C. (1990). Notes toward a neo-Luddite manifesto. *NASTS News*, *3*(3).

Harkness, J. (1992). Position paper on STS assessment and evaluation in K–12 schools. *NASTS News*, *5*(1), 7–8.

Henninger, E. H. (1991). Letter to the editor. *NASTS News*, *4*(1), 3, 10.

Jarcho, I., et al. (1990). Letter to the editor. *NASTS News*, *3*(4), 6.

Klein, J. T. (1990). *Interdisciplinarity: History, theory, and practice*. Detroit: Wayne State University Press.

Roy, R. (1991). Present efforts in technological literacy. In R. Jones (ed.), *Technology Literacy Workshop Proceedings*, (pp. 23–37). Newark: University of Delaware.

Shamos, M. (1990). An open letter to Irma Jarcho, John Roeder and Nancy Van Vraken (Teachers Clearing House for Science and Technology). *NASTS News*, *3*(6), 3–4.

Williams, W. F. (1990). To be critical or not to be—that is the question. *NASTS News*, *3*(6), 3–4.

Winner, L. (1986). *The whale and the reactor*. Chicago: University of Chicago Press.

CHAPTER 27

THE IOWA CHAUTAUQUA PROGRAM: A PROVEN IN-SERVICE MODEL FOR INTRODUCING STS IN K–12 CLASSROOMS

Susan M. Blunck
Robert E. Yager

THE ORIGINAL CHAUTAUQUA

The Iowa Chautauqua Program was created in 1983 when the National Science Teachers Association received major support from the National Science Foundation (NSF) to initiate a national Chautauqua program designed to improve kindergarten through twelfth-grade science teachers and courses. It was modeled after the highly successful Chautauqua Program of the American Association for the Advancement of Science project, which was also supported by NSF for two decades as a major effort to improve science teaching at the college level. Iowa was selected as one of seventeen states for the NSTA-NSF Chautauqua Program.

Basic to the Chautauqua model is the identification of new course titles-themes, nationally recognized instructors, and teachers and classrooms ready for some innovative practices. The program initially included only two short courses—one organized in the fall to share new ideas and approaches and one in the spring designed for the participants to share experiences, to synthesize the results, and to formulate a model for staff development. Such was the AAAS Chautauqua model, which was immensely popular and successful as a program to improve college science classrooms.

The Iowa Chautauqua Program for kindergarten through twelfth-grade teachers was established to introduce teachers to the emerging STS reforms that were receiving national attention from the NSF research effort called Project Synthesis. STS was one of the curriculum components headed by E. Joseph Piel (College of Engineering, Stony Brook). STS had been identified as a focus area for Harms's Project Synthesis (Harms and Yager, 1981). Dr. Piel was selected as the primary instructor for the Iowa Chautauqua effort, which operated as an NSTA-NSF effort during the 1983–1986 interim. He was the logical person to lead the effort since he was selected by Harms to head the STS focus group for Project Synthesis (Harms, 1977).

As with any new project, changes were adopted based on experiences from previous years. Initially the project included work with thirty teachers with fall awareness sessions concerning STS, a three- to five-week curriculum model tried in the classrooms of participating teachers, and the spring report and synthesis session. After one year the Iowa Utility Association (made up of all investor-owned utilities in the state) volunteered to be a cosponsor of the Iowa Chautauqua Program. This meant increasing the program to include five sites each year with thirty to fifty teachers enrolled. These five centers were rotated on a three-year cycle to be located in each of the fifteen Area Education Agency/Community College geographical regions of the state.

Other NSF funding and Eisenhower grants to higher education institutes in Iowa permitted a new element added to the Chautauqua model. A three-week experience with learning science in an STS format means that teachers can try a five-day module before the Fall Short Course. This meant more time could be spent on assistance with the use of instructional technologies and on assessment strategies prior to trying the STS strategies with any of their own students.

With 250 new teachers enrolled for the 1992–1993 academic year, the total number of teachers who have participated over a seven-year period numbers 1,700. These teachers have been employed in 283 of Iowa's 431 districts. A total of thirty of the most successful teachers are recruited each year to become Lead Teachers. They are enrolled in a two-week Leadership Conference in June as plans for three-week summer workshops and the short courses for the next academic year are finalized.

For the Iowa Chautauqua Program the NSTA descriptions of STS are used to characterize the STS modules, which are tried for four to six weeks in kindergarten through twelfth-grade classrooms (between the fall and spring short courses). These features include: (1) student identification of problems with local interest and impact; (2) the use of local resources (human and material) to locate information that can be used in problem resolution; (3) the active involvement of students in seeking information that can be applied to solve real-life problems; (4) the extension of learning going beyond the class period, the classroom, and the school; (5) a focus on the impact of science and technology on individual students; (6) a view that science content is more than concepts that exist for students to master on tests; (7) an emphasis on process skills that students can use in their own problem resolution; (8) an emphasis on career awareness—especially careers related to science and technology; (9) opportunities for students to experience citizenship roles as they attempt to resolve issues they have identified; (10) identification of ways that science and technology are likely to impact the future; and (11) some autonomy in the learning process (as individual issues are identified).

The STS classrooms are also seen as those that can be contrasted with textbook dominated classrooms. These contrasts include:

Textbook	*STS*
1. Textbook visible and used frequently	1. Textbook used only when it is needed as source of information
2. Teachers provide information for students to record and to repeat on tests	2. Teachers assist students in finding answers to their own questions; teachers rarely provide information (answers to student questions)
3. Activities are all prescribed, including goals, procedures, and often the results	3. Students plan activities as a way of testing their own ideas and explanations
4. Teachers rarely in communication with persons outside their own classrooms	4. Teachers utilize other teachers, parents, and experts in the community as sources for information and ideas
5. No focus on current problems and issues	5. Current problems and issues often provide the context for study
6. Science defined by what information is included in the textbook	6. Science defined as questions, possible answers to questions, and testing the possible answers that emerge
7. Teachers plan each lesson carefully	7. Teachers focus on goals and involve students in planning activities, actions, and sources of information
8. Teachers rarely admitting they do not know something arising in discussions	8. Teachers frequently admitting to not knowing; this stiuation is used to plan group actions to deal with an initial "not knowing" starting point
9. Student doing what text and teacher direct them to do	9. Students proposing actions, information sources, and new questions
10. Focus on words and terms from textbook	10. Terms rarely used as a focus by themselves; special terms are used only after meaning has been established

Textbook	*STS*
11. Science not viewed as operating in the school and/or community; that is, no local relevance	11. Nearly all questions, issues, and class activities have a base and a relevance at the local level
12. Ideas and information presented for mastery	12. Ideas and information sought out to respond to issues and questions
13. No use of newspapers and current periodicals	13. Frequent use of news reports and current situations
14. Much work on text and teacher-prepared worksheets	14. No work on text and teacher-prepared worksheets
15. Much time spent by teacher in preparing lessons	15. Students involved as much (if not more) than teacher in preparing for individual lessons
16. Class discussion and laboratories focus on competition and getting "right" answers	16. Discussion and laboratories focus on responding to issues, questions, and problems—often in a cooperative mode
17. Quizzes and tests focus on student recall	17. Evaluation focuses on what students can do; that is, how they can use information and skills
18. "Science" contained in place called science classroom or laboratory	18. Science in evidence in school as whole, the community, and the lives of students

The Iowa Chautauqua Model is summarized in Figure 27.1. The advantages that the program has shown to provide include:

1. awareness of a reform effort (STS) in a nonthreatening environment;
2. direct teacher experience with STS instruction (three-week summer workshop);
3. experienced STS teachers as vital parts of staff team;
4. assistance with planning a grade-level and school specific module;
5. encouragement to try the new module;
6. a focus on assessment strategies that are designed to provide evidence of the success (and failures) of new approaches;
7. a chance to compare results of STS teaching (Spring Short Course);
8. encouragement to communicate about the new efforts with STS (professional meetings, local staff development, and Leadership Training).

The effectiveness of the Iowa Chautauqua Model has been established by the Program Effectiveness Panel of the National Diffusion Network. The model

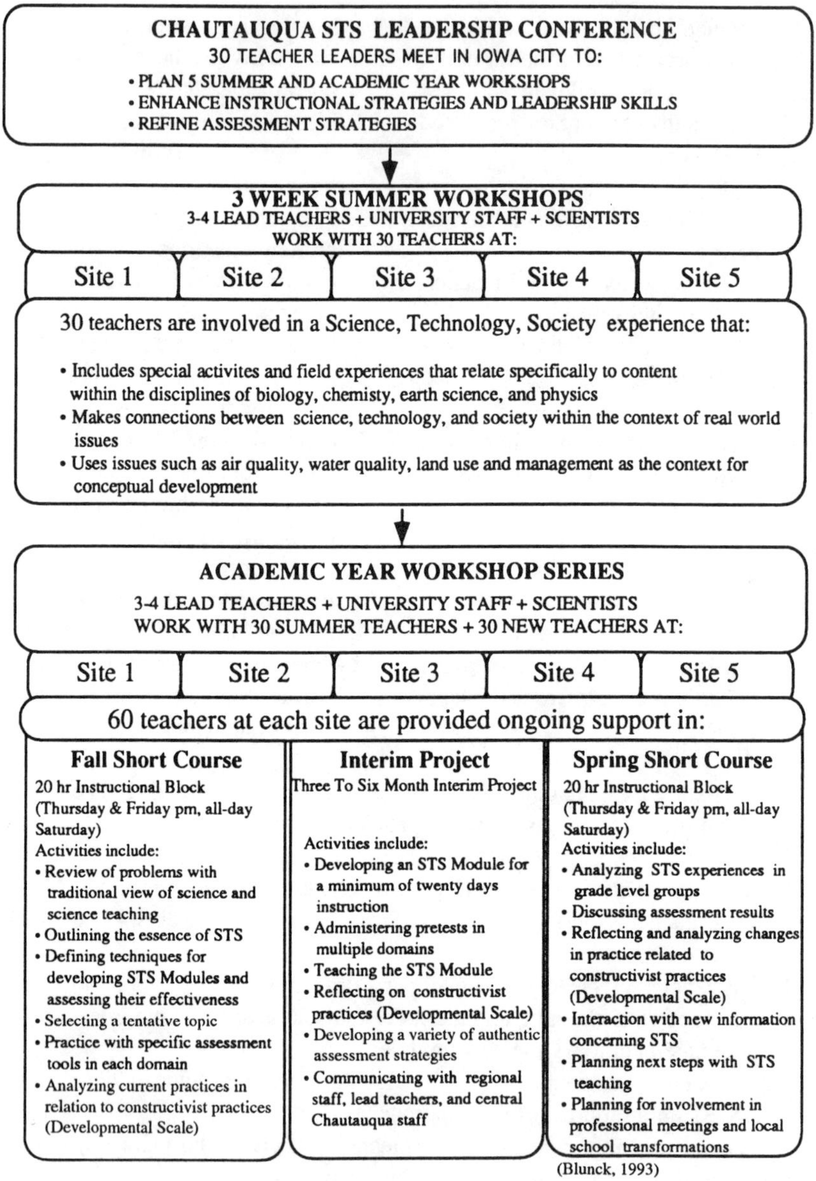

CHAUTAUQUA STS LEADERSHP CONFERENCE

30 TEACHER LEADERS MEET IN IOWA CITY TO:

- PLAN 5 SUMMER AND ACADEMIC YEAR WORKSHOPS
- ENHANCE INSTRUCTIONAL STRATEGIES AND LEADERSHIP SKILLS
- REFINE ASSESSMENT STRATEGIES

↓

3 WEEK SUMMER WORKSHOPS

3-4 LEAD TEACHERS + UNIVERSITY STAFF + SCIENTISTS

WORK WITH 30 TEACHERS AT:

Site 1	Site 2	Site 3	Site 4	Site 5

30 teachers are involved in a Science, Technology, Society experience that:

- Includes special activites and field experiences that relate specifically to content within the disciplines of biology, chemisty, earth science, and physics
- Makes connections between science, technology, and society within the context of real world issues
- Uses issues such as air quality, water quality, land use and management as the context for conceptual development

↓

ACADEMIC YEAR WORKSHOP SERIES

3-4 LEAD TEACHERS + UNIVERSITY STAFF + SCIENTISTS

WORK WITH 30 SUMMER TEACHERS + 30 NEW TEACHERS AT:

Site 1	Site 2	Site 3	Site 4	Site 5

60 teachers at each site are provided ongoing support in:

Fall Short Course

20 hr Instructional Block (Thursday & Friday pm, all-day Saturday)

Activities include:

- Review of problems with traditional view of science and science teaching
- Outlining the essence of STS
- Defining techniques for developing STS Modules and assessing their effectiveness
- Selecting a tentative topic
- Practice with specific assessment tools in each domain
- Analyzing current practices in relation to constructivist practices (Developmental Scale)

Interim Project

Three To Six Month Interim Project

Activities include:

- Developing an STS Module for a minimum of twenty days instruction
- Administering pretests in multiple domains
- Teaching the STS Module
- Reflecting on constructivist practices (Developmental Scale)
- Developing a variety of authentic assessment strategies
- Communicating with regional staff, lead teachers, and central Chautauqua staff

Spring Short Course

20 hr Instructional Block (Thursday & Friday pm, all-day Saturday)

Activities include:

- Analyzing STS experiences in grade level groups
- Discussing assessment results
- Reflecting and analyzing changes in practice related to constructivist practices (Developmental Scale)
- Interaction with new information concerning STS
- Planning next steps with STS teaching
- Planning for involvement in professional meetings and local school transformations

(Blunck, 1993)

FIGURE 27.1 The Iowa Chautauqua Model

can be disseminated to other states or smaller geographical areas to assist schools and teachers with reforms. Specific claims were verified as established by the Program Effectiveness Panel. They include:

I. The Chautauqua Program increases teacher confidence for teaching science.
   A. Planning science lessons.
   B. Involving students actively in learning.
   C. Matching goals with curriculum and instruction.
II. The Chautauqua Program increases teacher understanding and use of the basic features of science.
   A. Including focusing on questions.
   B. Including generating teacher- and student-generated explanations.
   C. Including teacher- and student-devised tests for determining the validity of explanations.
III. Lead Teachers involved with the Chautauqua Program are more able to stimulate their students to grow in six domains of science learning (compared to class sections where the techniques emphasized in the Chautauqua program are not used).
   A. A significant number of basic concepts are mastered (but not more than in traditional courses where the exclusive focus is on concept mastery).
   B. Students better understand basic processes of science.
   C. Students can apply concepts and processes to new situations.
   D. Students develop more creativity skills.
      1. Questioning.
      2. Proposing causes.
      3. Suggesting possible consequences.
   E. Student attitude is more positive following instruction.
      1. Toward science classes.
      2. Toward science teachers.
      3. Toward usefulness of science to them.
      4. Toward science careers.
   F. Students improve in their understanding of the basic features of science.
      1. Science means questioning, explaining, and testing.
      2. Science deals with activities that affect living in homes, schools, communities, and nations.
      3. Science is a human activity that involves acting on questions about the universe.

The evidence is abundant concerning the advantages of STS (Yager, 1989, 1990*a*; Yager, Myers, Blunck, and McComas, 1990). And, STS has been found to be most effective with female and minority students (Iskander, 1991; Liu, 1992; Lu, 1993; Mackinnu, 1991). The evidence is also clear that the Chautauqua is an exemplary inservice model correcting many of the problems noted with inservice education. These advantages include:

1. Planning, implementing, and evaluating involves all stakeholders; collaboration is evidenced for every aspect;
2. Inservice is well-planned and provides ongoing support; frequent opportunities for individual and collegial examination and reflection on instructional and institutional practices is provided;
3. All stakeholders help set clear and achievable goals;
4. Needs of teachers, school, and community are considered in setting realistic goals;
5. Inservice program is integral part of total school program providing teachers with ongoing support; a developmental approach is used;
6. A wide range of instructional strategies are used to accommodate teacher differences;
7. Instructional approaches focus on learning science content through the perspectives and methods of inquiry;
8. Theory, modeling, practice, and feedback are integral parts of the instructional process;
9. Choice is provided; teachers choose the experiences that best fit their needs;
10. Focus on changing teacher behavior;
11. Intrinsic and extrinsic incentives are part of the program, emphasis is placed on intrinsic rewards;
12. Positive leaders help create a shared vision and commit funds, time, materials, and human resources;
13. Administrative staff participates in the inservice;
14. Applications of adult learning principles are evidenced in all phases of the program;
15. Fostering a positive self-concept is the guiding force; professionals come to view themselves as life-long learners; and
16. Programs are ambitious and complex; evaluation is ongoing providing for continuous improvement.

References

Blunck, S. M. (1993). *Evaluating the effectiveness of the Iowa Chautauqua Program: Changing the reculturing behaviors of science teachers K–12*. Unpublished doctoral dissertation, University of Iowa, Iowa City.

Harms, N. C. (1977). *Project Synthesis: An interpretative consolidation of research identifying needs in natural science education.* (A proposal prepared for the National Science Foundation.) Boulder: University of Colorado.

Harms, N. C., and Yager, R. E. (eds.). (1981). *What research says to the science teacher*, Vol. 3. Washington, D.C.: National Science Teachers Association.

Iskandar, S. M. (1991). *An evaluation of the science-technology-society approach to science teaching*. Unpublished doctoral dissertation, University of Iowa, Iowa City.

Liu, C. (1992). *Evaluating the effectiveness of an inservice teacher education program: The Iowa Chautauqua program*. Unpublished doctoral dissertation, University of Iowa, Iowa City.

Lu, Y. (1993). *A study of the effectiveness of the science-technology-society approach to science teaching in the elementary school*. Unpublished doctoral dissertation, University of Iowa, Iowa City.

Mackinnu. (1991). *Comparison of learning outcomes between classes taught with a science-technology-society (STS) approach and a textbook oriented approach*. Unpublished doctoral dissertation, University of Iowa, Iowa City.

National Science Teachers Association. (1982). *Science-Technology-Society: Science education for the 1980s*. Position Paper. Washington, D.C.: Author.

Yager, R. E. (1989). Development of student creative skills: A quest for successful science education. *Creativity Research Journal*, *2*(3), 196–203.

Yager, R. E. (1990). Instructional outcomes change with STS. *Iowa Science Teachers Journal*, *27*(1), 2–20.

Yager, R. E., Myers, L. H., Blunck, S. M., and McComas, W. F. (1990). The Iowa Chautauqua program: What assessment results indicate about STS instruction. In D. W. Cheek (ed.), *Technology Literacy V: Proceedings of the Fifth National Technological Literacy Conference, Arlington, Virginia, February 2–4, 1990* (pp. 133–147). Columbus, Ohio: ERIC Clearinghouse for Science, Mathematics, and Environmental Education.

CHAPTER 28

NATIONAL SCIENCE EDUCATION STANDARDS: A TURKEY, A VALENTINE, OR A LEMON?

Janice Koch

THE "STANDARDS" MOVEMENT

The National Committee on Science Education Standards and Assessment first met in the spring of 1992 to consider the overwhelming task of developing National Standards for Science Education, addressing Curriculum, Teaching, and Assessment. With a grant from the U.S. Department of Education, the National Research Council agreed to take the lead to convene and coordinate a process that would lead to National Science Education Standards for kindergarten through twelfth grade. Dr. James Ebert, vice president of the National Academy of Science was designated as chair of the NCSESA to oversee both the development of science education standards and a nationwide critique and consensus process designed to produce the final draft by late 1994.

In a process determined to promote the outstanding practices of exemplary science teachers, the NCSESA embarked on their ambitious task by subdividing into three working groups for science curriculum standards, science teaching standards, and science assessment standards.

Representatives from the National Science Teachers Association, the American Association of Physics Teachers, the National Association of Biology Teachers, and the Earth Science Education Coalition joined representatives from the American Association for the Advancement of Science, the American Council of Science (ACS), and the Council of State Science Supervisors (CS^3) to ensure that the voices that permeated the final document were rooted in the rightful domain of the science teaching profession. While still a work in progress, the NCSESA has distributed different colored samplers of the materials that they are developing. The sampler of November 1992 was referred to as a "turkey," while the red-covered February 1993 sampler was called the "valentine." The yellow-colored July 1993 Progress Report could be thought of as the "lemon." New outlines that define eight "content" dimensions of science have been released twice during 1994. Such broad definitions for science exemplify the STS reforms.

The purpose of this essay is to review the salient features of the evolving National Science Education Standards and to explore those aspects of science education that appear, thus far, to be absent. The premise of this discussion is based on the fact that many children enter kindergarten as natural wonderers, curiously seeking to find out how and why their world works as it does. Beginning as many youngsters do with a healthy curiosity about the natural world, it is sad that conventional school science manages to squelch that excitement and teaches many youngsters to view science as arcane and unrelated to their own experience.

It is entirely possible that the new National Science Education Standards will refine and polish an already unproductive system of science education by ignoring the intrinsic importance of teaching science within the context of human experience.

The young child is an opportunist, grasping the moment when he or she can mess around, pull apart, explore, and invent. What Dewey referred to as "scientific habits of the mind" is the baggage of the young child, frequently questioning, wondering, seeking new answers. Rarely satisfied with an adult's response to a question and more eager to find out for him- or herself, the small child's world is filled with: But why? But how? Somewhere along the continuum of this small child's science education, the child learns not to question; he or she must do as the teacher suggests and not deviate from the instructions. The goal is to get it right—to be the first to do so whenever possible and to play the "teacher game." The rules are clearly elucidated by the teacher and the child who succeeds is the one who has integrated the rules of the game and follows instructions well.

Science education has not made space for the truly creative, inventive child who believes that there may be another way. Formal science education is deeply committed to prescribed content. It is the driving force in conventional science education, and it does not dare the student to think for him- or herself. Further, it has not accessed the scientific ways of being that many children employ as they interact with their environments. There is no provision in the content standards for validating the lived experience of students as they make their way through the world. In order to make science more accessible to our students we need to learn how to code much of their daily life experience as scientific. They classify, organize, experiment, draw conclusions, and employ trial and error frequently in their daily experience. For some children the very act of getting ready to arrive at school and the places and people they encounter along the way requires a set of thinking skills that enable them to negotiate their journey successfully.

Once, while doing science with middle school children on a small rural island off the coast of Maine, I learned how to spot a clam in the wet sand off this island coast. The middle schoolers who taught me how to spot a clam were adept at observing the slightest differences in the surface of the sand. The clues to their clam's whereabouts involved an almost instinctive ability to perceive minor changes in their coastal environment. Their skill in this arena is a valuable attribute for scientific study. Nobody ever told them that. We taught each other that day. As they learned more about the vegetation in their local environment, I learned how to spot a clam. We both understood the value of careful observations in noticing small differences in the flora and fauna that surrounded us.

Students come to us with profound personal knowledge of their own natural habitats. Science educators must look to that knowledge to build curriculum with and through the students' lived experience. The day after the San Francisco earthquake, I asked a student what he did in his earth science class. He said, "Page 72." Science education needs to help students to make meaning of their world. It cannot exist apart from the human contexts of personal, regional, and global issues.

Content Standards

The evolving standards suggest that "students should learn science in ways that reflect the modes of inquiry that scientists use" (National Committee on Science Education Standards and Assessment, 1993, p. 1). Do all scientists work in just one way? Is their work neat and organized, or can it be messy, intuitive, filled with second guessing, and frequently circuitous? Do ordinary thinking people use these modes of inquiry, or are they reserved just for scientists? Hazen and Trefil, in their book, *Science Matters* (1991), argue that "Real science like any human activity, tends to be a little messy around the edges" (p. xix). Conventional science education presents science as neat and linear with prescribed steps that scientists perform with rote precision and predictability. It denies the human side of science that remains tentative and unsure. Growing up in these rapidly changing, significantly confusing, and cumbersome times, it is difficult to relate to an area of study that is as deterministic as school science appears to be.

Hazen and Trefil remark that "to function as a citizen, you need to know a little biology, a little geology, a little physics, and so on. But schools are set up to teach one science at a time" (1991, p. xvii). While steps have been made through the NSTA's Scope, Sequence, and Coordination efforts to foster connections between the branches of science, there remains, entrenched in our system, a specialist mentality. This gives rise to the valuing of specialized knowledge as a starting point and what I perceive as the central theme of the evolving standards. Who decides what *all* students should know about science? There is an ongoing debate about preparing our students to function as scientifically literate citizens. What does "knowing science" mean? Trefil and Hazen make the distinction between *doing* science and *using* science. We need to make a commitment to all students, those who will want to *do* science for their life's work as well as all students who need to *use* science in their lives.

As part of the content standards, we need to talk about the term "fundamental understandings" in science. It does not automatically follow that developing fundamental scientific understandings will enable students to use this scientific knowledge to make meaning of many of the important newspaper items containing scientific developments. Anticipating this, the standards state that "teachers and school systems must stimulate every student to reach beyond these fundamental understandings and encounter additional thought-provoking, engaging science experiences that focus on local resources and environments, reflect personal and regional interests and expertise, and illustrate important uses and applications of fundamental understandings" (National Committee on Science Education Standards and Assessment, 1993, p. 5). While adding this statement has merit, it implies that the "making connections" part of using scientific knowledge has been relegated to the back of the bus. The driver remains that all powerful science content area. The passengers on the bus are "modes of inquiry" and "fundamental understandings" that all students should develop.

We have made a deity out of traditional cognitive knowledge in science. Well-meaning teachers seek to "cover" the science material and rarely think about embedding the content in a larger social and cultural context. This essay argues for placing the varied and variable personal and regional interests at the center of science education and watching the science emerge. The fear of placing the science subject matter at the heart of the standards cannot be overestimated. Despite good intentions and a broad perspective on science teaching, the NCSESA will reinforce traditional reliance on science as a body of knowledge to be "mastered." The Standards have the disturbing potential of being content prescriptive, reinforcing conventional modes of transmission that have evolved as the hallmark of science education.

Central to this critique is the belief that there are no real categories of separate science disciplines anymore and that the contexts in which scientific and technological understanding mediates responsible decision-making need to be central to the development of National Science Education Standards.

Often, it feels like there is a deep belief that modern science gives us the "true truth"—that, in fact, if only we can figure out an effective way to *give* that true truth to our students, they can join the ranks of the capable problem solvers of the universe and automatically understand how to apply fundamental scientific concepts in order to cope with the complexities of their world. By not addressing the teaching and learning of science through the various contexts of human activities in which individuals interface with science in everyday life (and in more complex technological applications through the processes and products of science and technology), the Science Education Standards are doomed to be read as one more content-driven mandate for science educators. Oh, and yes, after the content, be sure to make the connections, demonstrate science in human affairs, relate the activities to local events and resources, and foster decision-making about public issues related to science. The latter mandate feels adjunctive, an afterthought, tangential to the central theme of what all students should know in science. There are clearly articulated distinctions between content domains and curricula (National Committee on Science Education Standards and Assessment, 1993, p. 6), and equal importance is allotted to science as inquiry, science subject matter, scientific connections, and science and human affairs. However, conventional science curricula have been content driven in exactly the same three general domains of the natural science that are articulated in the July report: Science in Physical Systems, Science in Earth and Space Systems, and Science in Life Systems.

The July report indicates that "Depending on locally-elected options, science curricula in support of these fundamental understandings [in the three domains] can be organized in a variety of ways, ranging from interdisciplinary courses, issue-centered modules, spaced learning (SS&C), to more conventional formats" (National Committee on Science Education Standards and Assessment, 1993, p. 6). This statement provides an option for the reader, not a mandate.

There is a real fear among scientists and some science educators that approaching the teaching and learning of science through the contexts in which it presents itself to students will somehow "water-down" the pure science and render the study of science as "lesser." The irony, however, is that for many students, there is little lasting understanding of scientific concepts as they are currently presented, and even less ability to understand how these concepts present themselves in daily life. Nearly half of the graduating seniors at a prestigious private university were unable to correctly answer the question, "Why is it hotter in summer than in winter?" or "What is the difference between an atom and a molecule?" (Hazen & Trefil, 1991, p. 14).

The Science Education Standards provide us with the opportunity to stand conventional science on its head. Turning it upside down enables us to see the learners in all their complex cultures and socioeconomic strata. It could provide an impetus for real change in the way we think about conventional science education. As Robert Yager (1991) suggests, "Modern Science does not give us truth; it offers a way for us to interpret events of nature and to cope with the world." The following points reflect some of the thinking from Robert Yager and the Science, Technology and Society community of science educators. These need to be central to the articulation of the Science Education Standards if we are to create a new culture in science education:

1. The students need to be seen as knowers—as having their own ideas that are honored and validated by curricula. We need to seek out and use students' questions and ideas to guide whole instructional units.
2. The inquiry model needs to begin with the students' own questions and relate student thinking to the many ways in which science is practiced. The notion that scientific inquiry is solely practiced by scientists and exists apart from the task of daily living implies that we only have something to *give to* and very little to *access from* our students.
3. The fundamental understandings must be based on what science concepts students find useful in their own lives. These concepts need to be seen as a commodity for dealing with problems. Learning these concepts occurs because of a broad range of activities including, but not limited to, the science as inquiry model.
4. Students see science learning as a way to become involved in personal and social issues as they relate science studies to their daily lives. Scientific ways of thinking are fostered by group problem solving. Science process skills are seen as a refinement of skills that the students already possess and need to be engaged in developing more fully.
5. Students see science as a way of dealing with problems.
6. Science activities are developed as a means for helping students to answer their own questions and enabling groups of students to solve problems related to their local and global environments.
7. The study of science enables students to make meaning of their natural world and the technological developments that impact their lives.

Finally, the students should come to understand that the study of science is fun. Hazen and Trefil (1991) refer to it as, "just plain fun—not just good for you like some foul-tasting medicine. It grew out of observations of everyday experience by thousands of our ancestors, most of whom actually enjoyed what they were doing" (p. 19). Women were so intrigued with the study of natural science in the nineteenth century that they frequently attended women's colleges, studied natural science, and worked for nothing! Teachers need to model the sheer joy of finding solutions to problems, of making sense of some natural phenomena, and of using science to make decisions about events in our world that will shape the future of the planet.

TEACHING STANDARDS

The preliminary statements about the teaching standards imply that teachers will be expected to "select" science curriculum for their students that reflects the content standards and the range of ways that diverse students learn science (National Committee on Science Education Standards and Assessment, 1993). This notion of selection of curriculum implies that there are prescribed curricula that ensure strict adherence to the subject matter standards. There is no mention of the contributions of the students to the creation of science curriculum. Teaching standards need to include the teacher's role in helping the students to identify scientific problems in their own environments while facilitating the development of individual and group processes that enable students to construct solutions to their own problems.

The teaching standards need to address the collaborative nature of science teaching and learning. The science teacher needs to model what it looks like to be engaged with the wonders of natural science. The ability to be open to learning *with* and learning *from* the students is essential for successful science learning. Science teachers need to develop the skills to involve students in their own learning. The model of science teacher as science expert transmitting information in a cultural vacuum to unwitting, captive students is the dominant science teacher stereotype. The preliminary statements about the teaching standards include strategies that enable teachers to be "flexible and open to the ideas, strengths, and needs of students . . . [Teaching] Strategies must also enable students to construct their knowledge in a social environment and communicate their ideas in many forms" (National Committee on Science Education Standards and Assessment, 1993, p. 8). This is a hopeful component of the preliminary teaching standards; it beckons further elucidation.

ASSESSMENT STANDARDS

The assessment standards represent the strongest section of the July 1993 Progress document. They acknowledge the limitations of traditional science content assessments and the "current abuses of assessment information as a result of over-interpretation and over-generalization" (National Committee on Science Education Standards and Assessment, 1993, p. 12). The assessment standards elucidate the importance of assessing not only what students have learned but the *opportunity* to show what they have learned.

The assessment standards section of the Progress report broadens the ways in which assessment can be thought about. According to the document, the quality of assessment depends upon how well an instrument can measure student knowledge *and* what students can do with that knowledge. Without stating this explicitly, there is the implied critique of traditional testing that assesses student acquisition of discrete bits of scientific knowledge. This section supports "assessment strategies that are well matched to the intended uses of this information" (National Committee on Science Education Standards and Assessment, 1993, p. 12). The understanding that science assessment necessarily can take many forms is the underlying principle of this section. It is equally valuable that the assessment standards assert that there are important limitations to assessment information and that those limitations need to be made available to the users of science assessment data.

The assessment standards section identifies the importance of assessing "the full range of goals of science education" (National Committee on Science Education Standards and Assessment, 1993, p. 12). Assessing more than just knowledge of content is a major step toward changing the way scientific knowledge is viewed by teachers and students. Assessing the contexts in which scientific knowledge exists suggests that teaching those contexts is part of the Standards. This needs to be more fully developed in the final Content Standards.

The assessment standards elucidate the importance of the role of the students in participating in the formulation and construction of science assessments (National Committee on Science Education Standards and Assessment, 1993). The process of assessment can become a vehicle for further learning and another area of student growth. This standard also affirms the multiple ways of assessing the goals of science education, including formal and informal evaluations of students' work. This has important implications for the role of teachers in creating a sense of community in their science classes.

Further, the assessment standards address the importance of embedding assessment tasks in a variety of contexts, engaging students with diverse interests and backgrounds, making assessments fairer and more equitable. Acknowledging that assessment construction is a complex, multifaceted process that can serve different education needs is the spirit of the evolving assessment standards.

Leaving no stone unturned, the assessment standards have the potential to encourage important changes in science education practice and policy. As they have been elucidated thus far, they are forward thinking and a sign that these Science Education Standards may portend genuine transformation.

Programs and Policies Standards

Program standards will develop criteria for judging whether science programs provide all students with positive attitudes toward science; a foundation for understanding science in physical, earth and space, and life systems; opportunities to make meaningful connections among the sciences and between science and other spheres of human activity; and science programs that provide a community for science learning in which all students participate (National Committee on Science Education Standards and Assessment, 1993). The program standards are aligned with the subject matter standards that continue to describe science content within the traditional three categories. While it is necessary for the program standards to align themselves with the content, teaching, and assessment standards, the "connections in curriculum" component of the content standards may be regarded as an "extra" add-on with formal science content taking priority over the contexts in which science can be explored and understood.

Policy standards will develop criteria for judging alignment of: assessment practices with the Content and Assessment Standards and teacher preparation and certification with the Teaching Standards. These standards will also develop criteria for assessing the quality of communication within the science education system and between the science education system and the other systems with which it interacts. Further, these standards will develop criteria that policymakers can apply to assess the match between policies and resource allocation.

There is a critique and consensus feedback form attached to each of the National Science Education Standards documents thus far published and distributed. The November Sampler, the February Enhanced Sampler, and July Progress Report, and the September "semifinal" outline all invited responses from all members of the science education community. There is a strong commitment to creating a final document that reflects the interests and concerns of the diverse communities of science educators. It is in this spirit that this chapter was drafted. It is of real concern that the students' voices have representation within the final version of the National Science Education Standards. The understanding that the students bring diverse agendas to the science classroom needs to be accommodated within the context of these standards. What is missing, thus far, is the expressed need to place the science teaching and learning process within the context of the students' own experience. While the July document, in its introduction, expresses a commitment to *science for all* as a guiding principle, it does not explain how the content standards will address the needs of diverse learners. There appears to be a reliance on equating "high quality science experiences" with a system of delivery that does not access the lived experience of students. The notion that scientific principles and processes do emerge from diverse contexts of human experiences is absent. Although the Teaching Standards are not fully developed, they must recognize the importance of helping teachers to develop the strategies that will encourage their students to "initiate and explore questions, problems and ideas meaningful to them and applicable to their daily lives and interests" (National

Committee on Science Education Standards and Assessment, 1993, p. 9). The Content Standards address subject matter areas that promote teaching systems that *deliver* science information and scientific investigations. This is an inherent problem with the Science Education Standards. They do not acknowledge the science content that emerges from science teaching and learning that is driven by problem solving situations embedded in a social context, enabling the students to process the science information and make meaning of it for their world.

REFERENCES

Hazen, R. M., and Trefil, J. (1991). *Science matters: Achieving science literacy*. New York: Anchor Books/Doubleday.

National Committee on Science Education Standards and Assessment. (1993). *National science education standards: July '93 progress report*. Washington, D.C.: National Research Council.

Yager, R. (1991). The constructivist learning model: Towards real reform in science education. *The Science Teacher*, *58*(6), 52–57.

CHAPTER 29

STS MEANS SCIENCE ACROSS THE CURRICULUM: OR A WHITE COAT FOR EVERY TEACHER

John W. McLure

SCIENCE IS EVERYWHERE

This chapter takes the position that science needs to expand beyond the boundaries of its traditional turf. To the religious reader, this may sound like an evangelical message, and to the politician it could smell of imperialism. My motive is closer to the former than the latter, for I feel strongly that science is interesting, important, and fun, and I would like to help others embrace this view. I will contend that curriculum in much of the United States is in a state of imbalance, and that science can help strengthen educational programs and bring them into a harmonious state. This message should reach science teachers and the larger audience of experienced educators who do not yet think of themselves as science teachers.

Previous chapters have addressed in detail many issues and problems that face science educators and classroom teachers. I will give a brief synopsis of those that bear on this chapter.

First, we still walk in the hot ashes that fell after *A Nation at Risk* (National Commission on Excellence in Education, 1983) and the eruptions of dozens of other educational reform proposals. It is not easy to link cause and effect when viewing the impacts of these reports, yet we may conclude that the public was alarmed over what were perceived to be inadequate teaching and learning and that the country attempted to do something about the low scores of U.S. students in science and mathematics (U.S. Department of Education, National Center for Statistics, 1991). Many school districts have increased their years of science required for graduation to two years. College registrars in almost all the Big Ten universities further interpret two of those as coming from the Big Three—biology, chemistry, and physics. With this elitist interpretation in hand, secondary schools have made earth science a dumping ground. Grant moneys have summoned science teachers to summer institutes similar to the post-Sputnik pattern, and these veterans have updated themselves, shared their own ideas, and built curriculum together in an era that probably places a higher premium on constructivism than did the 1960s.

The legacy of the 1980s included a deemphasis, even a cutting back, into the disciplines that were not a part of the Big Five (science, mathematics, social studies, language arts, and computers). For teachers in art, music, physical education, and the practical arts such as industrial arts (or technology) and home economics (or domestic science), this era has been one long drought. Some teachers who are not educated in the Big Five have engaged in defensive credentialing to fit there, while those who have stayed on course often feel quite insecure. A third group of teachers have searched their souls, found new missions, and changed directions, as in the case of the industrial technologists. We remain in a state of curriculum imbalance that preceded the *A Nation at Risk* more than thirty years ago (National Commission on Excellence in Education, 1983; Tanner, 1971).

School Structures

Second, there has been dissatisfaction with school structures. International comparisons have harped on the limited exposure to physics that American students require in comparison to that in countries such as Russia and Japan. Current scope and sequence models have spent more time on fewer concepts, given more concrete instruction before moving to the abstract, and have dealt with each discipline each year. Science, mathematics, and geography will soon occupy elevated positions from which to view their disciplines based on their very recent scope and sequence efforts. They have taken seven-league steps toward the mapping out of the idealized content structures proposed by Bruner and others in the post-Sputnik era.

Some searchers for new structures involve an entire school. A small network of approximately thirty schools (fewer than that number initially) sponsored by the Association for Supervision and Curriculum Development (ASCD) explored a philosophical route toward restructuring. They reinvented student learning contracts and touched on division rearrangements of the sort that higher education has been trying out (Roberts & Cawelti, 1984).

A much larger consortium, Theodore Sizer's Coalition of Essential Schools, is still actively searching for the secondary school suitable for the next century (Goldberg, 1993).

The past thirty-plus years have witnessed the rise of the middle school in the United States. Many of these institutions have made brave attempts to revise the predominant high school structure, with strict departmental boundaries, which typified the junior high programs. The middle school faculty is organized into teaching teams, each with representatives of several disciplines. A team plans together and carries out counseling functions as well as academic ones. Because of the dynamism of these teams, the middle schools are stimulating places for faculty and students. Despite the high marks given to the middle school by the Carnegie Commission's 1988 *Turning Points* report and other claims to superiority, the middle school carries at least one flaw, in my opinion (Carnegie Corporation, 1989).

That blemish appears in these same, highly touted teaching teams. The popular composition includes a mixture of academicians dominated by the language arts, social studies, mathematics, and science teachers. Faculty from the practical and fine arts, foreign language, and physical education are classified as teachers of "exploratories" and shunted off to their own team. In the older junior high movement, the term exploratory usually meant a philosophical aim that was suffused throughout the curriculum. It applied to physical education as readily as it did to mathematics. There was a graceful egalitarianism about the term. The new use of exploratories suggests a collection of soft, hence vulnerable areas, a symptom of curriculum imbalance.

Site-based management offers one additional venture into restructuring. Journals like *Kappan* and *Education Leadership* have supplied us with a number of show-and-tell articles in the past few years on the success of this concept when applied in Middletown, U.S.A. It is interesting to speculate on what might happen in a school that combined site-based management with constructivism and W. Edward Deming's Total Quality Management (TQM) movement. The implications of such a combination for the science educator are intriguing (Deming, 1986).

Two last background issues at least remain to be mentioned—the flight from science, and the cry for interdisciplinary approaches to learning. The first of these, the flight from science, one usually encounters as "declining pupil interest in science," or words to that effect, in the writings of Stuart Yager and Bob Yager (1983–1984). One finds a litany of concerns over diminished enrollments in physics, over low high school student ratings of science in relation to other courses, and anguish over the low numbers and percentages of women and minorities who enter scientific studies in higher education and careers in the sciences. At the high school level, we wonder about gender and minority differences in science enrollments and achievement. Patricia White (1992) has documented the disturbing and persistent gap between male and female students' SAT scores, for example.

As one speaks of differences in achievement in science, it is easy to fall into the lazy habit of thinking of the discipline as a fixed body of knowledge. Paul DeHart Hurd and other keynoters have stressed the ever-increasing, mushrooming magnitude of scientific knowledge and the impossibility of keeping up with it.

How do we find paths to lead us through the swamp of knowledge so that we still manage to understand its habitat? State departments of education suggest that we reach for more interdisciplinary studies. Some of the accrediting associations, notably the North Central and Southern, are promoting new outcomes-based assessments that emphasize only a few interdisciplinary goals, such as better communication, as the basis for periodic evaluations. Curriculum developers push the "infusion" approach for the interweaving of various movements into the traditional liberal arts courses. Among these infused movements are multicultural, career, and health education (Fogarty, 1991).

Field theories may provide another way to avoid suffocating the students under mountains of new facts. Reflect for a moment on the reasons why some science writers stand the test of time. Who are they? Loren Eiseley and Aldo Leopold of previous generations, and Steven Jay Gould, John McPhee, Lewis Thomas, and Annie Dillard of the recent scene, to name a few. Part of their success must be due to their rich, strong, prose, and another part to the ease with which they entertain us with their simple explanations of esoteric phenomena. Surely also their works appeal because they interrelate so many aspects of life, large and small. One can glimpse plant-animal interactions, evolution at work, plate tectonics and mountain building, the rise and fall of communities, cycles, extraterrestrial influences, and wine in old bottles in one chapter or essay by these writers.

There is a certain attraction to the idea of interdisciplinary studies. Each of us goes through a French encyclopedist phase now and then. Yet we must admit that the other side of the coin exists. An Israeli science educator, Nir Orion, is convinced that interdisciplinary approaches cause students to lose their focus on geological field trips. Independently, a geologist in the Missouri geological survey, Art Hebrank, arrived at the same conclusion. When Art leads a workshop for schoolteachers in the largely igneous and mineral rich St. Francois Mountains of the Eastern Ozarks, he resists the pleas of some teachers for a stop at a quarry to collect fossils because he feels that such a foray into paleontology diverts the focus away from his chief subject (Brophy and Alleman, 1991).

Many specialists will agree with Orion and Hebrank. Interdisciplinary studies, humanities courses, and similar blends are stimulating, they say, but the real path lies with specialization. Perhaps science educators work in a crossroads between these two positions, the interdisciplinary and the specialized. It may be that they have something to offer their colleagues in other disciplines from this vantage point. I will say more about this in the development of the solution to science across the curriculum.

A Needed Movement: Science across the Curriculum

What is "science across the curriculum," and why do we need it? The answer to the first part of the question depends on what one's definition of science is. When language arts teachers launched the "reading across the curriculum movement" fifteen or so years ago, they seem not to have tried to define reading, instead, they identified skills and strategies that the disciplines could use. The Iowa reading model included parallel steps, with some differences in the substeps for each discipline. These steps were: (1) selecting reading materials, (2) planning assignments, (3) introducing reading materials, (4) guiding student reading, and (5) eliciting and evaluating students' responses to reading.

The Iowa guide for reading across the content areas contained some suggestions for science that sound as if they were already a part of the usual classroom practice, such as having a useful knowledge of "course goals, objectives, and a basic sequence of subject matter," and assessment of the difficulties of vocabulary for various students.

What then did the reading teachers offer the science teachers that seemed to represent particular aids from the language arts? Probably the best answer lies in the overall five-stem design mentioned above. In addition, the reading movement introduced Cloze procedure, readability formulas, and the Skimming, Questioning, Reading, Recalling, and Reviewing (SQ3R) reading strategy. In other sections the guide mentioned concepts and skills such as reading load, mapping, and the analysis of style. They included the following patterns of writing: (1) the generalization-elaboration pattern, (2) sequence, or sequential style, (3) comparison-contrast, (4) cause and effect, (5) problem statement (where necessary information for the solution of a problem is stated sentence by sentence), and 6) induction (observations or events are followed by a hypothesized explanation that is then elaborated) (Conner, et al., 1980).

When language arts teachers promoted the reading across the curriculum movement, they designed it as an infusion effort, not a major competing unit of time. They recognized that reading was already to some extent an act infused throughout the curriculum; yet teachers needed assistance to help their students succeed with the reading material.

Science may not be so obviously pervasive in the disciplines as is the act of reading, depending on what we mean by science. George Gaylord Simpson provided this definition: "Science is an exploration of the material universe in order to seek orderly explanations (generalizable knowledge) of the objects and events encountered—*but these explanations must be testable*" (1969, p. 82).

Simpson's definition may be compared with some of the essential skills among the reading teachers. One imagines a scientist gathering data to test a hypothesis, analyze the results, and accept or reject the hypothesis. Science includes additional mental operations, and these need to be identified and exported to other disciplines, even though they may be shared goals already. Consider some ways of thinking that students of science use. Let us see whether they may be applicable to other places in the curriculum.

One may begin with perception and mindsets. A mycologist, Robert Embree, says that his beginning students show a perceptual mindset in their fieldwork: they search only on the ground for fungi; they don't look up, above their heads, along tree trunks. He pushes his students to break their mindsets about where to expect to find fungi. And when one does find them, the discovery may jolt the senses. The specimen may contain subtle colors that surpass the ability of the field guide to represent. Neither the key nor the field guide quite prepares one for the robust taste of hen-of-the-woods (*Grifola frondosa*) or the pungent and very agreeable odor of the elm-tree mushroom (*Hypsizygus tesselatus*). At a later point in his course, Embree asks his students to design their own key for the identification of a group of fungi. In this brief example, there are three operations for the science educator to mull over for possible export across the curriculum. They are: (1) teaching students to break mindset and shift to another perceptual channel, (2) science as aesthetic experience, and (3) an elementary exercise in taxonomy.

Some science educators have good advice and useful protocols to offer to help students work upward through Piagetian stages of development. Darrell Phillips has been a leader in this area for decades. A teacher who has studied under Phillips and attended to Piaget and Kohlberg carefully may see *Alice in Wonderland* for the first time as a young character who passes from a child who does what she is expected to do in the beginning of the story to a more autonomous girl capable of handling abstract ideas when she finally defies the Queen of Hearts.

One contribution that a science educator can make to colleagues across the curriculum is to show reverence for precision. This attitude is not unique to the field of science, certainly; a masterful English instructor will revise a passage ten times if necessary, and many a band member has gone home at night with tired lips from one more rehearsal to make "The March of the Sirdar" come out just right. Even so, there is an indelible lesson for some students who see the care that a scientist takes in collecting and recording data and specimens. Few can help but be impressed by the tenacity of a bird-watching friend of mine on a field trip. In a scene typical of her, we were near the shore of a backwater. Hundreds of swallows roosted and flew about among low trees, mostly willows. These were not unusual birds. If we had seen peregrine falcons, everyone would have insisted on the exact number. Because of their almost continual movement, it appeared to be impossible to form more than an approximate number for the entire flock of swallows. But Carole stumbled over roots, skidded on silt, and clawed through the undergrowth till she reached a far more accurate count than any of the rest of our group. That was a lesson in precision.

From ways of thinking, let us turn to methods and content that a science educator may export across the curriculum. Science, Technology, and Society deserves mention here. The key characteristics of this movement for me are (1) its reliance upon issues and problems, (2) the hands-on uses of everyday science, chemistry in the kitchen, physics in the garage, and (3) the opportunity that it gives to students to construct their own meanings.

In the same buildings where STS has appeared in science classes, some industrial arts educators have renamed their field "industrial technology," and apparently by accident they are often revising their course content so that it overlaps with STS activities. One finds units on transportation, robotics, energy, flight, and solar cooking, for example, in middle school industrial technology classes. Surely, here is a place and time for science teachers to join hands with their industrial technology colleagues. By doing so, they may provide more applications and reinforcement for their STS. With very little effort, science and industrial technology (and perhaps domestic science) teachers can work together in the same teaching teams that mark the restructuring of the middle schools, avoid the second class citizenship that the exploratories have come to mean, and help to restore some balance to the secondary curriculum.

STS provides what Susan Blunck, Caroline Giles, and Julia McArthur call an "inclusionary approach to science teaching." Prior to doing their research, they were impressed by the book, *Women's Ways of Knowing*. The book describes women scientists as "constructed knowers" for whom "connections between science and human beings were a very important concern." Men and women readers have said that the description matched their experiences with science (Belenky, Clinchy, Goldberger, and Tarule, 1986).

Blunck and Ajam (1991) looked at gender-related attitudes toward science and science teachers. Before the STS treatment, female learners showed more negative attitudes toward science than their male peers. After the instruction, the attitudes of the females shifted significantly in a positive direction. The gap between males and females was greatly narrowed (Blunck, Giles, and McArthur, 1993). Their findings tend to support the position of Marcia Linn and Janet Hyde (1989) that the gender differences in mathematics and science are declining, "that small gender differences in cognitive and psychosocial domains be deemphasized and instead that learning and earning environments be redesigned to promote gender equity."

Although the hands-on activities in STS attract teachers and students, there are possibilities for strengthening connections in textbooks. Suppose we would like to infuse more science into social studies. Let us examine the science that appears in two recent leading textbooks of American history and speculate on what changes we might recommend.

The first textbook, by Joseph R. Conlin, *Our Land, Our Time*, in two volumes, by Coronado, appears to be written for grades seven through twelve (Conlin, 1986). The second, John A. Garraty's *The Story of America*, in two volumes, a Holt, Rinehart and Winston/HBJ work, seems to be designed for the high school reader (Garraty, 1992).

The beginnings of both books mention glaciation and the Bering land bridge. The topic of glaciation serves to introduce early man to the New World. A few pages afterward, Conlin speaks of Mayan medicine, though he does not describe it. Garraty mentions the atlatl with nothing about its effectiveness.

The authors settle into a pattern of predominantly political narrative with periodic lists of technological advancements: indigo dye, the cotton gin, the steel plow, and the radio. There is almost no mention of the science that led to the inventions.

Conlin and Garraty rely on their sister discipline geography (and not geology) occasionally to help students interpret mountains, basins, bays, and fall lines. Events like the gold rush of 1849 and the discovery of oil in Pennsylvania are interpreted primarily for their social impacts.

Garraty includes a rare full page on two women Nobel Prize winners, Rosalyn S. Yalow for her work on radio immunoassay and Barbara McClintock for her studies of "jumping genes." Conlin includes a paragraph explanation of carbon 14. Overall, the two history textbooks list inventions, tell of society's uses of technology, and say very little about science.

Summary and Recommendations

The field of science offers multiple skills, ways of thinking, aesthetic moments, methods and contact that can strengthen other fields, help solve some of our educational problems, and restore a degree of balance in the curriculum. The time is appropriate for science educators to join with teachers in other disciplines in the formation of teaching teams and other structures that may enhance instruction. Science consultants and supervisors should take the lead in setting up state guidelines for the export of science across the curriculum. We need competing sets of guidelines to spark this movement rather than a single, monolithic, nationwide model.

References

Belenky, M. F., Clinchy, B. M., Goldberger, N. R., and Tarule, J. M. (1986). *Women's ways of knowing*. New York: Basic Books.

Blunck, S. M., and Ajam, M. (1991). Gender-related differences in students' attitude with STS instruction. *Chautauqua Notes*, *6*(2), 2–3.

Blunck, S. M., Giles, C. S., and McArthur, J. M. (1993). Gender differences in the science classroom: STS bridging the gap. In R. Yager (ed.), *What research says to the science teacher: Volume 7, The science, technology, society movement* (pp. 153–169). Washington, D.C.: National Science Teachers Association.

Brophy, J., and Alleman, J. (1991). A caveat: Curriculum integration isn't always a good idea. *Educational Leadership, 49*(2), 66.

Carnegie Corporation. (1989). *Turning points: Preparing American youth for the twenty-first century.* The report of the Task Force on Education for Young Adolescents. New York: Author.

Conlin, J. R. (1986). *Our land, our time.* (2 Vols.) San Diego: Coronado.

Conner, J., et al. (1980). *Guides for teaching secondary students to read in subject areas.* Des Moines: Iowa Department of Public Instruction.

Deming, W. E. (1986). *Out of the crisis.* Cambridge: Massachusetts Institute of Technology.

Fogarty R. (1991). Ten ways to integrate curriculum. *Educational Leadership, 49*(2), 61–65.

Garraty, J. A. (1992). *The story of America* (2 Vols.). Austin, Tex.: Holt, Rinehart & Winston (HBJ).

Goldberg, M. F. (1993). A portrait of Ted Sizer. *Educational Leadership, 51*(1), 53–56.

Linn, M. C., and Hyde, J. S. (1989). Gender, mathematics, and science. *Educational Researcher, 18*(8), 17.

National Commission on Excellence in Education. (1983). *A nation at risk: The imperative for educational reform.* Washington, D.C.: U.S. Department of Education.

Roberts, A. D., and Cawelti, G. (1984). *Redefining general education in the American high school.* Alexandria, Va.: Association for Supervision and Curriculum Development.

Simpson, G. G. (1969). Biology and the nature of science. *Science, 139*, 81–88.

Tanner, D. (1971). *Secondary curriculum: Theory and development.* New York: Macmillan.

U.S. Department of Education, National Center for Statistics. (1991). *The condition of education, 1991, Volume 1, Elementary and secondary education* (pp. 38–39). Washington, D.C.: Author.

White, P. E. (1992). Women and minorities in science and engineering: An update. Washington, D.C.: National Science Foundation.

Yager, S. O., and Yager, R. E. (1983–84). Perceptions of science of third, seventh, and eleventh grade students enrolled in Cedar Rapids (Iowa) schools. *Iowa Science Teachers Journal, 20*(3), 9–14.

Endword

This volume offers the challenge to the 1981 report of Project Synthesis. The authors end the report calling for seven actions including:

1. A major redefinition and reformulation of goals for science education; a new rationale, a new focus, a new statement of purpose are needed. These new goals must take into account the fact that students today will soon be operating as adults in a society that is even more technologically oriented than at present; they will be participating as citizens in important science-related societal decisions. Almost total concern for the academic preparation goal, as is currently the case, is a limiting view of school science.
2. A new conceptualization of the science curriculum to meet new goals; redesigns of courses, course sequences/articulation, and discipline alliances are needed. The new curricula should include components of science not currently defined and/or used in school. Direct student experiences, technology, personal and societal concerns should be foci.
3. New programs and procedures for the preparation, certification, assignment, and the continuing education of teachers; planned changes, continuing growth, and systems for peer support are needed. With new goals and a new conceptualization of the science curriculum, teachers must have assistance if their meaning is to be internalized. Without attention to inservice education, new directions and new views of the curriculum cannot succeed.
4. New materials to exemplify new philosophy, new curriculum structure, new teacher strategies; exemplars of the new directions, that is, specific materials for use with learners, are constantly needed. They provide concrete examples for use in moving in such new directions.
5. A means for translating new research findings into programs for affecting practice; a profession must have a philosophic basis, a research base, a means for changes to occur based on new information. Separation of researcher from practitioner is a major problem in science education: all facets of the profession must work in concert for major progress to occur.

6. Renewed attention to the significance of evaluation in science education; self-assessment strategies, questioning attitudes, massing evidence for reaching decisions on instruction and student outcomes are basic needs. Without such questions, observations, and judgments, future changes will be merely haphazard occurrences.
7. Much greater attention to development of systems for implementation and support for exemplary teaching and programs at the local level; current erosion of support systems for stimulating change and improvement in science education at all levels is a major problem. (Harms and Yager, 1981, pp. 129–130)

STS provides a framework, examples, and evidence of what has happened for each of these needs more than a decade later.

However, Harms's closing comments reported elsewhere in this volume warrant repetition: "Not only is there an increased need to understand large national issues, there is also an increasing need to understand the way science and technology affect us as individuals. Thus, a new challenge for science education emerges. The question is this: 'Can we shift our goals, programs and practices from the current overwhelming emphasis on academic preparation for science careers for a few students to an emphasis on preparing all students to grapple successfully with science and technology in their own, everyday lives, as well as to participate knowledgeably in the important science-related decisions our country will have to make in the future?'" (Harms and Yager, 1981, p. 119). Even with the emerging research relative to STS, there remain many who fear change, who do not or cannot examine the evidence, who continue as if there is no crisis.

STS provides the most complete body of research and experience for any of the current reforms in the United States However, Voelker's analysis remains a current challenge: "If we want a science program that is truly responsive and responsible to the citizen in a scientifically and technologically oriented society, we must elevate current and future citizen concerns. We cannot assume that curricula which emphasize traditional cognitive knowledge and an understanding of the scientific process will lead to an understanding of the science-related issues confronting society. Neither can we assume that such traditional curricula will assist our student-citizens in applying their scientific knowledge and processes to these issues. Some sacred cows of the science curriculum must be eliminated. But the short-term trauma this sacrifice may elicit will be replaced by a long-term gain for all citizens" (Yager, 1982, p. 79).

The full promise of STS will not be realized without continuing trial, experimentation, analysis, discussion, and action. This volume is offered to encourage all of these.

REFERENCES

Harms, N. C., and Yager, R. E. (eds.). (1981). *What research says to the science teacher*, Vol. 3. Washington, D.C.: National Science Teachers Association.

Yager, R. E. (ed.). (1982). *What research says to the science teacher*, Vol. 4. Washington, D.C.: National Science Teachers Association.

INDEX